ADVANCES IN HUMAN ECOLOGY

Volume 6 • 1997

EDITORIAL ADVISORS

ADVANCES IN HUMAN ECOLOGY

Editor: **LEE FREESE**
Department of Sociology
Washington State University

VOLUME 6 • 1997

 **JAI PRESS INC.**

Greenwich, Connecticut   *London, England*

CONTENTS

LIST OF CONTRIBUTORS

Sandra G. F. Bukkens
National Institute of Nutrition
Rome

Sing C. Chew
College of the Behavioral and
Social Sciences
Humboldt State University

C. Dyke
Department of Philosophy
Temple University

Raúl García-Barrios
Centro de Investigacion y Docencia
Económicas, A.C
México D.F.

Mario Giampietro
National Institute of Nutrition
Rome

Ulrich Müller-Herold
Department of Environmental
Sciences
Swiss Federal Institute of Technology

David Pimentel
Department of Entomology
Cornell University

Rolf P. Sieferle
Historical Institute
University of Mannheim

Rod Swenson
Center for the Ecological Study
of Perception and Action
University of Connecticut

Peter J. Taylor
Center for Critical Analysis
of Contemporary Culture
Rutgers University

Adam S. Weinberg
Department of Sociology
and Anthropology
Colgate University

PREFACE

EDITORIAL POLICY

This series publishes original theoretical, empirical, and review papers on scientific human ecology. Human ecology is interpreted to include structural and functional patterns and changes in human social organization and sociocultural behavior as these may be affected by, interdependent with, or identical to changes in ecosystemic, evolutionary, or ethological processes, factors, or mechanisms. Three degrees of scope are included in this interpretation: (1) the adaptation of sociocultural forces to bioecological forces, (2) the interactions between sociocultural and bioecological forces, and (3) the integration of sociocultural with bioecological forces.

The goal of the series is to promote the growth of human ecology as a transdisciplinary problem-solving paradigm. Contributions are solicited without regard for particular theoretical, methodological, or disciplinary orthodoxies and may range across ecological anthropology, socioecology, biosociology, environmental sociology, ecological economics, ecological demography, ecological geography, epidemiology, and other relevant fields of specialization. The editor will be especially receptive to contributions that promote the growth of general scientific theory in human ecology. No single volume will represent the full range the series is intended to cover.

CONTENTS OF VOLUME 6

This volume may be the most multidisciplinary so far published in the series. The subjects of these papers range, in order, from ecological interpretations of physical principles, through evolutionary and environmental connections, and into social organization.

Rod Swenson presents an argument that relates the second law of thermodynamics to self-organizing systems. Swenson's purpose is to show, in contrast to traditional interpretations, that physical systems can be expected to produce order not disorder; and so, the production of order in biological systems is neither anomolous nor serendipitous. With this result established, classic Cartesian dualisms that separate the dead from the living and humans from their environments are shown to be invalid.

C. Dyke contrasts the theoretical imagery of traditional science with the imagery now emerging from the sciences of complexity. Dyke, like Swenson before him, argues that the new theoretical imagery is much more suited to the interactive dynamics that define human ecology. Uses of the new imagery and disuses of the old are illustrated for the framing of macroeconomic theory.

In accord with the new theoretical imagery described by Swenson and Dyke, Mario Giampietro develops a formal theory of self-organization, grounded in energetics, to characterize societal development and evolution. A principal result of modeling the theory for socioeconomic systems shows how technological changes imply trade-offs between short-term efficiency and long-term adaptability such that both can not be simultaneously maximized. The companion paper that follows, by Giampietro, Sandra Bukkens, and David Pimentel, develops six formal applications.

Ulrich Müller-Herold and Rolf Sieferle provide a theoretical description of a dynamic evolutionary pattern, called a risk spiral, by which changing conditions of subsistence production and its associated surplus change the terms of human survival. The changed terms are theorized to change the adoption of strategies by which ruin is avoided and risk is prevented or managed, with new conditions for risk unfolding as new types of uncertainty are introduced from strategies previously adopted.

Sing C. Chew provides historical documentation on how ecological processes of degradation—especially land degradation as a consequence of deforestation—affect societal processes such as capital accumulation, economic growth, and the intensification and decline of political organization. Chew provides a detailed examination of data over long historical periods concerning the production, consumption, and distribution of timber and wood products as evidence for these societal-environmental connections.

Peter Taylor and Raúl García-Barrios challenge neo-Malthusian environmentalist views of biophysical limits to population growth, and the rhetorics of the views, as they are conventionally interpreted for the environmental degradation found in some developing nations. The authors explore an alternative theory frame, sociologically grounded, which ties environmental degradation to conditions that promote poverty and undermine the institutional sufficiency of indigenous peoples to effectively adapt.

Adam S. Weinberg presents qualitative research on grassroots environmental groups, the data from which enable him to show how environmental law can become an obstacle to environmental reform and can exacerbate environmental conflict. Weinberg argues that law frames the experience and, thus, conditions the strategies employed for the mobilization of grassroots organizations, thereby limiting the structure and process by which environmental conflicts may be resolved.

Lee Freese
Editor

AUTOCATAKINETICS, EVOLUTION, AND THE LAW OF MAXIMUM ENTROPY PRODUCTION:
A PRINCIPLED FOUNDATION TOWARD THE STUDY OF HUMAN ECOLOGY

Rod Swenson

ABSTRACT

Ecological science addresses the relations between livings things and their environments, and the study of human ecology addresses the particular case of humans. However, there is an opposing tradition built into the foundations of modern science, which separates living things and particularly humans from their environments. This tradition, with its dualisms traceable from Descartes through Kant into Darwinism with its grounding in Boltzmannian thermodynamics, precludes a truly ecological science. A deeper understanding of thermodynamic law and the principles of self-organizing (autocatakinetic) systems provides the nomological basis for dissolving

Advances in Human Ecology, Volume 6, pages 1-47.

Cartesian incommensurability, for putting evolution back in its universal context, and for showing the reciprocal relation between living things and their environments, and thereby provides a principled foundation for eco-logical science in general and human ecology in particular.

INTRODUCTION

The word ecology was coined by Haeckel and used in his Generelle Morphologie in 1866 to refer to the science of the relations between living things and their environments (Bramwell 1989), and by this general definition, human-environment relations constitute the central subject of human ecology. The idea of the separation of humans from their environments, however, is deeply embedded in the foundations of modern science. Descartes, promoting a psychology versus physics dualism, where the active, epistemic part of the world (human "minds") was incommensurably separated from what was taken to be the dead, mechanical, physical part of the world ("matter" or "other"), provided the world view that became the basis of modern science and which, at the same time, supernaturally separated humans from the world (see also Dyke 1997, this volume).

Later, arguing that the active, end-directed striving of living things in general could not be accounted for within the dead, mechanical world of physics, Kant, calling for the autonomy of biology from physics, promoted a second major dualism, between biology and physics, or between living things in general (not just human minds) and their environments (Swenson and Turvey 1991). The Cartesian tradition was carried into evolutionary theory with the ascendancy of Darwinism which, making no use of physics in its theory, provided an explanatory framework where "organisms and environments," in Lewontin's (1992, p. 108) words, "were totally separated." Strong apparent scientific justification for these postulates of incommensurability came with Boltzmann's view of the second law of thermodynamics (the entropy law) as a law of disorder—a hypothesis that he developed during the last quarter of the last century in an attempt to save the Cartesian, or mechanical, world view.

According to Boltzmann, physical systems are expected to become increasingly disordered or run down with time, and the spontaneous transformation of disordered to ordered states is "infinitely improbable" (Boltzmann 1974 [1886], p. 20). This view effectively set the active nature of living things—as expressed, for example, in the fecundity principle, perhaps the sine qua non of Darwinian theory (the idea that life acts to produce as much biological order as it can), and in the progressive ordering that characterizes the evolution of life on Earth as a whole (from bacterial eco-systems some four billion years ago to the rise of civilizations and the global proliferation of culture going on today)—against the apparent, otherwise universal, laws of physics. The world, in this view, was supposed to be running down according to the laws of physics, but biological and cultural systems seemed to be about "running up"—to be not about going from more orderly to less orderly states but about producing as much order as possible. It is "no surprise," under these circumstances, in the words of Levins and Lewontin (1985, p. 19), "that evolutionists [came to] believe organic evolution to be a negation of physical evolution." As Fisher (1958 [1930], p. 39), one of the founders of neo-Darwinism, expressed it, "entropy changes lead to a progressive disorganization of the physical world…while evolutionary changes [produce] progressively higher organization." This view is still at the foundations of the Darwinian view today, as evidenced by Dennett's (1995, p. 69) definition of living things as things that "defy" the second law of thermodynamics.

Cartesian incommensurability precludes an ecological science. Consequently, ecological science, if it is to be about what it purports to be about—living thing/environment relations—requires a theory that dissolves it. The postulates of incommensurability came into modern science on the issue of the active, epistemic dimension of the world, and this is precisely the battleground where they must be defeated. In particular, the confrontation must occur at the interface of physics, psychology, and biology, and the distinguishing characteristic of this interface is that it is defined by intentional dynamics, the dynamics that, not coincidentally for ecological science, distinguishes the living thing/environment relation. By intentional dynamics, I refer to end-directed behavior prospectively controlled or determined by meaning, or "information about" (of which "end-in-mind" behavior is a lately evolved kind).

Rivers flowing down slopes or heat flowing down temperature gradients from hot to cold, are examples of end-directed systems, but they are not examples of intentional dynamics because they do not require meaningful relations to determine the paths to their ends. Their behavior is explicable in terms of local energy potentials and fundamental physical laws. In contrast, when a bacterium swims up a concentration gradient, a bird flies above the Earth or opens its wings to effect a landing on a branch, a human drives a car, or puts a satellite in orbit around the Earth, or moves some food from her plate to her mouth, this behavior is seen to go in directions

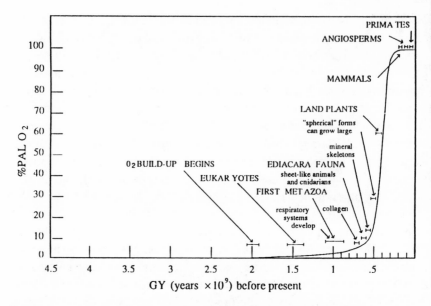

Source: From Swenson (1989a, p. 71). Copyright © 1989 by IEEE. Reprinted by permission.

Figure 1. The production of progressively higher states of order as a function of increasing levels of atmospheric O_2 in geological time (PAL is present atmospheric level.) Atmospheric oxygen was put into the atmosphere by life and has been maintained at present levels for some hundreds of millions of years by life at the planetary level. The transformation of the Earth's atmosphere from oxygenless to oxygen-rich, as well as the forms, including human cultural systems, that have systematically arisen as a consequence of it, are measures of the terrestrial system's departure from thermodynamic equilibrium, or progressive ordering. This runs counter to the widespread conception of the second law of thermodynamics due to Boltzmann, which predicts that the world should be becoming increasingly disordered. This has led evolutionists to believe that biological and cultural evolution defy or negate physical evolution—a belief, in effect, of two incommensurable "rivers," the river of physics which flows down to disorder, and the river of biology, psychology, and culture which flows up.

that are different, and often opposite, from those that follow causally from local physical potentials and laws. This kind of end-directed behavior, the kind that is meaningfully or epistemically determined with respect to non-local potentials, characterizes intentional dynamics. Terrestrial evolution shows the world to be in the order-production business, characterized not by progressive disordering to equilibrium but by the production of increasingly higher states of order, and the way the production of this "river that flows uphill" takes place—to use Calvin's (1986) felicitous phrase—is through meaningful, or epistemic, relations (see Figure 1).

To continue the metaphor and summarize in different terms what has been said above, if ecological science is to be ecological in more than name only, it must provide a principled basis for unifying what are otherwise taken to be two incommensurable rivers: the river of physics that flows downhill, and the river of biology, psychology, and culture that flows uphill. The absence of such a principled account invites the otherwise recurrent problem of the pre-Socratic Parmenides, who had a fully coherent theory of the world which, however, could neither account for nor even accommodate his own existence. Recent advances in the theory of thermodynamics and self-organizing systems provide the basis for dissolving the postulates of incommensurability, by providing the nomological basis for intention and intension in a physical world otherwise taken to be collapsing to disorder and to be inherently meaningless (defined exhaustively by extension). Rather than being anomalous, with respect to somehow defying or going against universal laws, the intentional dynamics of living things are seen to be a direct manifestation of them. This provides a principled basis for setting the active ordering that characterizes the evolution of life, from Archean prokaryotes to the present rapid globalization of culture in its universal context, and in so doing provides a principled foundation for ecological science in general and human ecology in particular.

THE CARTESIAN CIRCLE AND THE FIRST POSTULATE OF INCOMMENSURABILITY

The influence of Descartes, whose ideas were built into modern science at its origins, is hard to overestimate. Although the physics of Newton eclipsed the physics of Descartes, it was the latter's dualistic metaphysics that provided the ground on which the former was able to flourish; and, because psychology and physics were defined at their modern origins by Descartes, he is often referred to not only as the father of modern philosophy but as the father of modern psychology and physics as well. What Cartesianism effected with its dead, mechanical, or clockwork world view was a means for the religious authority of Descartes' time to interpret science within a context it could accept, and for humans to see themselves, in the words of Descartes (1986 [1637], p. 67), as "masters and possessors of nature." Humans, as privileged creations on Earth, were taken to be sitting dualistically outside the clocklike world, learning the laws of physics in order to manipulate them toward their own taken-to-be divine ends. In this view, there was no theory of cultural ordering, or of evolution in general, because humans and the static mechanical physical world they were said to inhabit were taken to have been created full blown by divine act.

Defining the Epistemic Dimension Out of the Physical World

A fundamental point to make with respect to the Cartesian world view is that by defining physics and psychology by their mutual exclusivity—call this the "first postulate of incommensurability" (Swenson 1996)—it literally defined the active epistemic dimension out of the physical part of the world altogether. According to Descartes, the world was said to be divided into an active, purposive, perceiving "mind" (the "free soul," "thinking I," "Cartesian ego," or "self") on the one hand (the psychological part), and passive, "dead," purposeless "matter" (the physical part) on the other. The physical part, defined exclusively by its extension in space and time, was seen to consist of reversible, quality-less, inert particles governed by deterministic causal laws from which the striving mind, seen as active, boundless, and without spatial or temporal dimension, was said to be immune. An immediate implication of this view was that spontaneous ordering in general, and intentionality and meaning in particular, were thus eliminated from the physical world by definition, and so needed to be extra-physically imposed from the outside. For Newton, Boyle, and other believers in Descartes' mechanical world view, who took the world to be extra-physically ordered by God, this was not a problem. It was a reaffirmation of their belief.

The Problem of Dualist Interactionism

Even if such a world were extra-physically given, there is an insurmountable problem with respect to how such a system could ever possibly work. This was recognized almost immediately by many of Descartes' own followers in his own time. This is the problem of dualist interactionism. In particular, if psychology and physics ("mind" and "matter," or "self" and "other") are dualistically defined the way Descartes did by their mutual exclusivity, then there is no way in fact that they could ever interact. Leibniz recognized this central problem of Cartesianism by anticipating the law of energy conservation (the first law of thermodynamics). For one thing to interact with another, he argued, requires something conserved over the interaction, and if something is conserved over the two things or processes, at some level they are part of the same thing. "There must be something which changes, and something which remains unchanged," wrote Leibniz (1953 [1714], p. 27)—anticipating, it could be argued, the second law of thermodynamics, too. Without some conservation, the two would be truly incommensurable—two separate worlds without any possible relation or causal connection.[1]

This separateness of the physical and mental was reinforced by Descartes' theory of perception and his famous cogito ergo sum: that what is known indubitably is the self-reflective mind perceiving itself. For Descartes, the indubitability of matter was not so clear. With mind ultimately perceiving itself (and the physical world exhaustively defined by extension and so excluded from the category of mental or meaningful things), Descartes' strong claim as to what is known or what might exist therefore did not include an "outside" world at all. The epistemic dimension of the world on Cartesian principles, it was soon realized, became a closed "Cartesian circle" with no way in or out. With immaterial mind perceiving itself and no grounds to assert meaningful relations with, or even the existence of, anything outside the individual self, ego, or self-motivating, self-reflective mind—in effect no environment at all, the active epistemic act, the subjective, was simply given. Such a view is clearly inimical to a theory of ecological relations.

Closed-Circle Theory, Cultural Ordering, and the Epistemic Dimension

It is not surprising that post-Cartesian theories of knowledge, intentionality, or meaning would become linked, explicitly or not, with theories of culture and evolution. Culture is clearly an epistemic process effected by meaningful relations, and the epistemic process itself would clearly seem to be evolutionary. What is interesting, however, is that post-Cartesian theories of knowledge are typically seen to be allied either with cultural or with evolutionary accounts as two competing paradigms—the work of the later Wittgenstein, Kuhn, and others being exemplars of the first, and that of Popper, Campbell, Lorenz, and others being exemplars of the second (Munz 1985, 1987). Supporters of the first view ("closed-circle theorists"), who have worn incommensurability and relativism almost as a kind of badge of enlightenment, look to sociology or social psychology as the basis for meaning and intentionality, while evolutionary epistemologists, supporters of the second view, look to evolutionary theory or, more particularly, to Darwinian theory as the ground for the epistemic dimension. In this subsection, I briefly review the former. The latter is discussed in the context of the next section, which deals specifically with evolution.

The roots of closed-circle theory can be found in Durkheim and Malinowski, in the "sociology of knowledge" of Mannheim, in Marx and Engels' earlier work on ideology, and in Spencer's work before them. All of these, however, should not be construed as closed-circle theorists in the extreme postmodern sense of Wittgenstein and Kuhn. The common thread that unites this lineage is that cultural ordering is seen to determine individual action. This core idea was later associated with what came to be known as functionalism, in contrast to what is sometimes called psychologism, the idea that cultural systems are rational constructions of individual intentional agents. In the former view, rather than culture being taken as the rational construction of individuals, instead individuals are taken as component productions of cultural systems. The contributions of the functionalists were substantial in that they recognized cultural systems as self-organizing systems. The problem, however, was that they had no theory of self-organization.

Malinowski, in explicit reaction to psychologism as well as the then-prevalent evolutionary views of history or culture, held that cultural systems were effectively closed circles where the parts all function to maintain the whole. Given that, in this view, the circular relations that define the system are seen to refer back to themselves—that the function of the system is to maintain itself—cultural systems were said to exist sui generis. Everything is thus explained with respect to something else that happens internal to the circular relations of the system. Here we see the beginnings of the transposition of the Cartesian circle from the individual to the cultural or social psychological level.

Wittgenstein took this latent idea of cultural Cartesianism and made it more explicit. The epistemic dimension of the world, rather than constituted through the self-referential circular relations of the individual human mind as it had been for Descartes, was said to be constituted through the intersubjective circular relations of humans within a cultural system. Meanings, said Wittgenstein, are formulated and stated in "language games" consisting of a set of rules that constitute closed circles of meanings. There are no individual meanings because there is no individual language, and because such systems are closed circles, there can be no ostensive pointing or reference to anything outside the system (i.e., an objective "world"). What is more, because meaning is entirely relative to the rules of each system and, thus, meaning invariance across cultural systems is denied, such circles of meaning are incommensurable with respect to each other. Truth thus varies from one closed circle to the next and can only be measured with respect to the rules, or authority, of a particular community.

In the influential history and philosophy of science of Kuhn, Wittgenstein's closed-circle language games were turned into paradigms, and the history of science was now seen as the shift from one paradigm to another (scientific revolutions). Because reality, according to this view, is taken to be the ideal construction of human cognizers operating under particular paradigms and since paradigms as closed circles are incommensurable with each other, there is no way to talk of progress in science, a direction in time, or advancement from one paradigm shift or revolution to the next. Without meaning invariance there is no way to make a comparison. In this view, Einstein's physics, for example, do not subsume or explain Newton's but are simply different. Neither one is "truer" than the other. They are simply incommensurable. The postmodern structuralism of Foucault and Derrida, and the postmodern pragmatism of Rorty which in effect uses Foucault to justify Wittgenstein (Munz 1987), are all closed-circle theories that share the common premises of the relativity of meaning to circularly closed systems and the incommensurability of such systems with respect to each other and an external world. Closed-circle theory carries forward the anti-realist position of positivism, but at the same time challenges its rationality.

While closed-circle theory is often given as a kind of enlightened alternative to modernism, it is itself modernism carried to a certain post-Humean, post-Kantian, extreme conclusion. The Cartesian core is still there, only wrapped in sociological packaging and transposed from the individual to the cultural level. The most severe problems fall into three main areas.

1.   Closed-circle theory is anti-evolutionary.

Because closed circles are incommensurable with respect to each other, there is no way to assert that they are part of an evolutionary process or that any such process even exists. There is no way to provide an ordinal measure with respect to time. Closed-circle theory is time-symmetric. From the view of closed-circle theory, Einstein's theory could have preceded Newton's; the theory of oxygen could have preceded the theory of phlogiston; the theory of heat and the conservation of energy could have preceded the caloric; and the periodic table of elements might just as well have come before the theory that earth, fire, air, and water constituted the basic elements. Closed-circle theory thus fails (or does not care) to recognize or account for evolutionary dynamics, and this includes the active and expansive nature of the epistemic act, or epistemic dimension, itself. This anti-evolutionary foundation is underpinned by the intersubjective idealism of closed-circle theory, which extends to the extreme the Cartesian-Kantian anti-ecological tradition of effectively putting humans at the center of the universe.

2.   Closed-circle theory invokes an illegitimate teleology at its core.

By making the fundamental reality the circular relations that define a cultural system, closed-circle theory, including its functionalist ancestors, substitutes formal causality (the form or shape of a thing—in this case, the circular relations) for the efficient cause that constitutes the usual notion of causality in modern science (e.g., such as that found in various bottom-up rationalist schemes such as "psychologism," or in billiard-ball mechanical models in physics). Cultural systems are seen to be self-organizing systems of sorts which, in the production of their components or component relations, function toward their own ends, in particular, to maintain themselves. But at the ground of modern science from which it starts and for which no replacement theory is offered, there is no principled basis provided for where such ends or end-directed behavior can come from. The ends simply point back to themselves, and this is precisely the problem that discredited virtually every one of closed-circle theory's functionalist ancestors before it (Swenson 1990; Turner and Maryanski 1979). The teleology of closed-circle theory is thus more of a kind of religious than a scientific assertion. It requires defeating some widely held scientific assumptions but provides no principled basis for doing so.

Downward causality has traditionally been rejected by biology because it does not fit into the explanatory framework of natural selection (discussed more fully below), and by physics because downward causality constitutes macroscopic ordering in a world which, according to the received view of thermodynamics, should be collapsing to microscopic disorder. In addition, no matter how it is assumed that closed circles get ordered in the first place, the fact that they remain so sui generis, or without outside relations or ostensive pointing, makes them ideal perpetual motion machines of the second kind—a flight in the face of what many (e.g., Eddington 1928) have called the most fundamental and unbreakable of all the laws of physics.

3. The intersubjectivity at the core of closed-circle theory begs the old Cartesian questions and doubles the problem.

Briefly put, meanings for the closed-circle theorist exist in the persistent and invariant relations constituted through the intersubjectivity that define the closed circle. To each individual, however, this requires persistent and invariant relations with a world *outside* herself or himself, and that requires a non-Cartesian theory of perception. In short, the intersubjectivity of closed-circle theory requires breaking the Cartesian circle at the individual level, since the individual mind is no longer simply perceiving itself but is perceiving something external, in relation to which it comes to be determined or defined. This requires a commensurability between knower and known which undercuts the ground of closed-circle theory. Once the individual Cartesian circle is broken, there is no principled basis to maintain the cultural one (viz., once one has admitted the fundamental existence of a self-other relation, there is no principled basis to confine this only to other humans).

EVOLUTIONARY EPISTEMOLOGY, ECOLOGICAL SCIENCE, AND THE PROBLEM(S) WITH DARWINISM AS THE THEORY OF EVOLUTION

The Second Postulate of Incommensurability

As noted above, Cartesian metaphysics came full-blown into modern biology with Kant, who argued correctly that the active striving of living things could not be fathomed as part of a dead, reversible mechanical world. Rather than questioning the impoverished physics, however, Kant promoted a second major dualism, between biology and physics, or between living things and their environments. Call this the "second postulate of incommensurability" (Swenson 1996). The argument, grounded on the view of the incommensurability between the active, striving, intentional dynamics of living things and their "dead" environments, is still promoted today by leading proponents of Darwinian theory (e.g., Mayr 1985). Boltzmann's interpretation or hypothesis of the second law of thermodynamics has played a crucial role, as already noted, in giving apparent legitimacy to the view that physics has nothing to say to biology—its principles being not simply foreign but hostile to it. Darwinian theory, from Darwin on, had little use for physics in its theory. Darwin, in Lewontin's words, "completely rejected [the] world view...that what was outside and what was inside were part of the same whole system" (1992, p. 108). This carried the anti-ecological Cartesian-Kantian postulates directly into evolutionary theory and made the theory as inimical to ecological science as its ancestral relatives.

Evolutionary epistemologists, as noted in the preceding section, have a view almost directly opposite to that of the closed-circle theorists (e.g., see Callebaut and Pinxten 1987; Radnitzky and Bartley 1987). Whereas closed-circle theorists such as Wittgenstein and Kuhn are arch anti-evolutionists, evolutionary epistemologists look to evolutionary theory, in particular to Darwinian theory, to provide an account of the epistemic dimension. Evolution, on this view, is taken to be a continuous and progressive knowledge acquisition process following from natural selection, in Popper's words, from amoeba to man. Every living thing, according to this view, has knowledge in the expectations on which its intentional behavior depends, and this knowledge, as a consequence of natural selection, is taken to be (hypothetically) true, since if not true, to put it simply, the living thing in question would be dead. While to the closed-circle theorist, true knowledge follows from cultural authority under a particular paradigm, to the evolutionary epistemologist it is determined with respect to the performance of an epistemic agent in the world. Scientific knowledge is seen to be continuous with evolution by natural selection, since it too involves a trial and error process of selection through the proposal and refutation of falsifiable hypotheses (Campbell 1987).

The problem with evolutionary epistemology is its reliance on Darwinian theory. Darwinism's Cartesian postulates eliminate it a priori from the task that evolutionary epistemologists would like to have it perform. More specifically, two immediate problems, either of which by itself would be sufficient to disqualify Darwinian theory from providing an account of the epistemic dimension, can be quickly given. They are mentioned here but are discussed in more detail with the other "big" problems of evolution below. The first is that Darwinian theory assumes intentional dynamics to begin with, and this puts an explanation of intentional dynamics outside its theory. The second is that the claim that evolution is a progressive knowledge acquisition process is an assertion that can be neither made nor explained on the grounds of Darwinian theory, because the relevant observable (fitness) is relativized to members of breeding populations. These and the other problems below can all be seen to follow from the position evolutionary theory has backed itself into as a consequence of the Cartesian postulates at its core.

General versus Specific Theories of Evolution

The dream of uniting the two apparently opposing rivers, it should be noted, did not escape Fisher, who imagined that the two apparently opposing directions of biology and physics "may ultimately be absorbed by a more general principle" (1958 [1930], p. 39). Lorenz, one of the founders of evolutionary epistemology, wrote that the aspect of life "most in need of explanation, is that, in apparent contradiction to the laws of probability, it seems to develop from...the more probable to the less probable, from systems of lower order to systems of higher order" (1973, p. 20). For Spencer (e.g., 1852, 1862, 1892), who defined the term evolution and popularized the idea in numerous best-selling books prior to Darwin, biological evolution was part of a more general universal process of evolution. Spencer defined evolution as a process of the transformation of less-ordered to more-ordered states following from natural law (the "law of evolution"). Spencer was never able to supply the physical basis for his law of evolution. As a consequence of its asserted, if not demonstrated, nomological continuity (viz., biological ordering as a special case of universal ordering), Spencer's general theory of evolution was at least an attempt at a commensurable rather than incommensurable theory and stands now as an early statement of evolution as a law-based self-organizing process.

With the ascendancy of Darwinism, evolution was taken out of its universal context, and the meaning of the term was reduced to biological evolution alone (see also Swenson 1991b, 1992, 1996, In press-a, In press-b). According to Mayr (1980, p. 12), the "almost universally adopted definition of evolution [today] is a change of gene frequencies" following from natural selection. This was the "final implementation" of the basic Darwinian concept, except that the focus was shifted by neo-Darwinism from organisms to genes. It was with the reduction of the meaning of the term evolution from a universal to a biological process that the Cartesian-Kantian postulates were built into the core of evolutionary discourse, and with them the major anomalies of Darwinian theory. These are not simply the problems of evolution but true anomalies with respect to Darwinian theory because, as will be seen, they are problems that its core postulates preclude it from answering.

The Problem(s) with Darwinism as the Theory of Evolution

There are six main problems with Darwinism to be highlighted and discussed:

1. Natural selection requires the intentional dynamics of living things
    in order to work, and this puts the intentional dynamics of living
    things outside the explanatory framework of Darwinian theory.

The core explanatory concept of Darwinian theory in all its various
forms is natural selection (Depew and Weber 1995). Evolution, according
to Darwinism, follows from natural selection, and natural selection is
entailed by a situational logic (Popper 1985): *If* certain conditions hold,
then natural selection will necessarily follow. These conditions are: herita-
ble variation, finite resources, and the fecundity principle, a biological
principle that captures the active striving of living things. Natural selection, said Darwin, follows from a population of replicating or reproducing
entities with variation "striving to seize on every unoccupied or less well
occupied space in the economy of nature" (1937 [1859], p. 152). Because
"every organic being" is "striving its utmost to increase, there is therefore
the strongest possible power tending to make each site support as much life
as possible" (Darwin 1937 [1859], p. 266). Paraphrasing Darwin, the
fecundity principle, which refers to the intentional dynamics of living
things, thus says that nature acts in a way that "maximizes the amount of
life per unit area" (Schweber 1985, p. 38) given the constraints. But notice
that the situational logic from which natural selection follows makes natural selection dependent on the intentional dynamics of living things. Natural selection does not explain the intentional dynamics; it is a consequence
of them, and this puts intentional dynamics outside the explanatory framework of Darwinian theory.

2. Darwinism has no observables by which it can address or account
    for the directed nature of evolution.

That evolution is a progressive or directed process (meaning, going in a
direction) is seen in the cited statements of Fisher and Lorenz and is evident
to anyone who looks at the planetary evolutionary record (e.g., see Figure
1). It is a core idea for evolutionary epistemology, which sees evolution as
a progressive knowledge acquisition process, as Popper put it, from
"amoeba to Einstein," where the knowledge a thing has is measured by its
"fitness." But Darwinism, in effect, is a time-symmetric theory and has no
observables that can be used to measure the direction of evolution at all,
especially fitness. Because fitness is relativized to members of breeding
populations, the fitnesses of different kinds of things, as in the case with

closed circles in closed-circle theory, are incommensurable with respect to each other and cannot be compared (e.g., Fisher 1958 [1930]; Sober 1984). One zebra that runs faster than another, better avoids predators and thus produces more offspring, can be said to be more fit than the slower zebra, but a zebra can not be compared on the same basis to a mouse or an amoeba. Mice can only be judged more or less fit than other mice, and amoebas with respect to other amoebas, and this makes fitness an incommensurable observable with respect to evolution writ large. Darwinian theory has no ground from which to measure or account for the directed nature of evolution and, in particular, no ground for evolutionary epistemology to claim evolution as a progressive knowledge acquisition process. To justify this claim would require evolution to be about something other than fitness.

3. Because natural selection works on a competitive population of many, and the earth as a planetary system evolves as a population of one, Darwinian theory can neither recognize nor address this planetary evolution.

One of the most important empirical facts that has come to be recognized in recent decades is that the Earth at the planetary level evolves as a single global entity (e.g., Cloud 1988; Margulis and Lovelock 1974; Schwartzman et al. 1994; Swenson and Turvey 1991; Vernadsky 1986 [1929]). The present oxygen-rich atmosphere, put in place and maintained by life over geological time, is perhaps the most obvious prima facie evidence for the existence and persistence of planetary evolution (see Figure 1). With the shift of the Earth's redox state to oxidative some two billion years ago, evolution undeniably became a coherent planetary process. Because the evolution, development, and persistence of all higher-order life has depended and continues to depend on the prior existence and persistence of evolution at the planetary level, this single planetary system may well be considered the fundamental unit of terrestrial evolution. Without question, an understanding of planetary evolution is fundamental to evolutionary theory, to ecological science, and to a theory of cultural evolution and human ecology. Yet, this poses a major problem for Darwinian theory because the planetary system as a whole cannot, by definition, be considered to be a unit of Darwinian evolution (Dawkins 1982; Maynard Smith 1988). Darwinian theory, which defines evolution as the consequence of natural selection acting on a competitive replicating or reproducing population of many, cannot address or even recognize planetary evolution because there is no replicating or reproducing population of competing Earth systems on which natural selection can act. The Earth evolves as a population of one. Natural selection is seen to be a process internal to the evolution of the planetary system and, thus, rather than explaining terrestrial evolution, natural selection awaits an explanation of planetary evolution by which it, as a manifestation, might be explained (Swenson 1991a).

4. Darwinian theory has no account of the insensitivity to initial conditions (like consequents from unlike antecedents) required to account for the reliability of intentional dynamics or the evolutionary record writ large.

Contemporary Darwinian theory is characterized by a commitment to the assumptions of gradualism, continuous change, reductionism, and efficient or mechanical cause. The dynamics of its theory are based on the difficult (if not impossible) marriage of a kind of Laplacean determinism, namely, that like antecedents produce like consequents—that, for example, if the initial conditions or microconditions are changed, the macroscopic dynamics will be different—and the belief at the same time that there exists a certain amount of microscopic randomness, variation, or "error" in the world. The latter is supported by the most widely held views of quantum mechanics (viz., that probability is, in fact, objective). The consequence of these assumptions with respect to terrestrial evolution writ large is that it is seen as a process where, in effect, "anything goes." Given the condition of microscopic randomness, if one rewound the tape of evolutionary history back to some point in the distant past and played it again, it would turn out "entirely different" every time one rewound the tape (e.g., Gould 1989; Williams 1992, p. 3). Yet, if such a micro-macro relation were true, if living things were sensitive to initial conditions in this way, the characteristic properties of terrestrial evolution writ large and, in particular, the intentional dynamics of living things, would be inconceivable. Real-world systems of this kind show a remarkable insensitivity to initial conditions: They are "end-specific" not "start-specific," to use Dyke's (1997, this volume) felicitous terms. They repeatedly produce the same end states from different initial conditions, and they are required to do so in order to survive, because, regardless of the ultimate facts of quantum mechanics, real-world initial conditions are never the same twice. This remarkable insensitivity to initial conditions on which terrestrial evolution as we know it depends, is unrecognized and unaccounted for by Darwinian theory.

5. The incommensurability between biology and physics assumed by Darwinian theory provides no basis within the theory according to which epistemic or meaningful relations between living things and their environments can take place.

The fecundity principle on which the Darwinian view of evolution crucially depends assumes the active intentional dynamics of living things— the meaningful determination of their end-directed behavior. However, given the Cartesian psychology or theory of perception at the core of Darwinian theory, the rejection by Darwinism that what is inside and outside are part of the same whole system (Lewontin 1992), there is no principled basis for meaningful relations to take place. The outside or physical world is a world of extension, while the inside world, the biological or psychological part, is a world of intension. This re-creates the Cartesian problem of dualist interactionism. An ecological science requires an *ecological* evolutionary theory, and such a theory requires a non-Cartesian theory of perception, or an ecological psychology, to show a principled basis according to which meaningful relations can take place. A theory such as Darwinism that holds biology and physics, or living things and their environments, to be incommensurable cannot provide a principled basis for meaningful relations; and, because the evolution of life is distinguished by intentional dynamics or meaningful relations, such a theory is deficient not only as an evolutionary epistemology and an ecological science but as a theory of evolution in general, too.

6. Evolution according to Darwinism is defined as a change in gene frequencies, and this puts cultural evolution outside the reach of Darwinian theory.

Clearly, as a consequence of the rate at which it is transforming the planet, cultural evolution is of great import to those interested in terrestrial evolution in general and ecological science in particular. For evolutionary epistemologists, cultural evolution is part of a continuous process of knowledge acquisition, and for human ecology, cultural evolution is clearly central to its subject matter. By defining evolution as a change in gene frequencies, however, Darwinian theory can have little to say about cultural evolution at all, which "is not really evolution at all" (Dawkins 1986, p. 216) under this definition. This is not a mere technical point. The interests of genes, and the interests of "memes" (roughly speaking, the ideas that are replicated by cultural systems as their principle hereditary component [Dawkins 1986; Dennett, 1995]) are incommensurable, and so are biological and cultural evolution on this view.

AUTOCATAKINETICS: A THEORY OF EMBEDDED CIRCLES

Symmetry Breaking and Symmetry Making: Autocatakinesis, and the Generalized Metabolism of Dynamic Flow Structures

An ecological science requires a demonstration of why, contrary to what most evolutionary theorists believe, biological and cultural evolution are not a negation of physical evolution. It requires a principled basis for uniting the two rivers, or otherwise apparently two-directional universe, which Fisher and many others have pointed out. It requires answering the Lorenz question about why evolution as a whole appears to be a progressive process that moves from more probable to increasingly less probable states. It needs to show why, if the transition from disorder to order is infinitely improbable, as Boltzmann argued, the world is in the order-production business. What is more, it must show the basis for the meaningful relations by which the intentional dynamics of biological and cultural ordering are distinguished.

Identity through Flow

As noted briefly above, part of the attraction of Descartes' passive, "dead," quality-less world of physics was that it required extraphysical ordering to get it ordered. The mechanical world, made of inert, reversible particles incapable of ordering themselves—as Boyle (Lange 1950 [1877], p. 255) pointed out, like the "ingenious clock of Strasburg Cathedral"—must have an intelligent artificer to account for it. In addition to Boyle, the argument from design was made repeatedly throughout the rise of modern science. Paley's famous version about finding a watch on a beach and knowing that it had to have had a watchmaker to design it, is the one Darwin is credited with undermining by using the idea of natural selection—from which came Dawkins' (1986) metaphor of the blind watchmaker. But there is a serious category error in these arguments—namely, that non-artifactual systems, such as living ones, are not the same kinds of things as mechanical artifacts. In different terms, if you found a watch on a beach, or wherever, it certainly would make sense to imagine that it had an artificer to design it, because nothing like it has ever been found in the universe, as far as anyone knows, that was not artifactually produced.

Machines or artifacts are defined by static order. Their identity is constituted and maintained by static components—the same components, external repairs excluded, in the same positions with respect to each other. Living systems, from bacteria to cultural systems, as self-organizing or spontaneously ordered systems, are defined by dynamic order. Their identities are constituted through the incessant flux of their components, which are continuously being replaced from raw materials in their environments and being expelled in a more dissipated form. Persistence (the form of the thing) at one level (the "macro" level) is constituted by change at the component level (the "micro" level). In more technical terms, living systems are autocatakinetic systems while artifactual systems are not. The class of autocatakinetic systems includes more than just living systems, and this immediately suggests a connection between living and non-living things that will become more apparent later.

Dust devils, hurricanes, and tornadoes, for example, are all autocatakinetic flow structures whose identities are constituted in just this way: by the incessant flux of matter and energy pulled in from, and then excreted or expelled back into, their environments in a more degraded or dissipated form (see Figure 2). An autocatakinetic system is defined as one that

maintains its "self" as an entity constituted by, and empirically traceable to, a set of nonlinear (circularly causal) relations through the dissipation or breakdown of field (environmental) potentials (or resources) in the continuous coordinated motion of its components (from auto-"self" + cata-"down" + kinetic, "of the motion of material bodies and the forces and energy associated therewith," from kinein, "to cause to move") (Swenson 1991a).

The importance of understanding living systems as flow structures with behavior generic to the class was emphasized in the first half of this century by Bertalanffy (e.g., 1952), and later by Schröedinger (1945), who popularized the idea of living things as streams of order which, like flames, constitute themselves by feeding off "negentropy" (energy potentials) in their environments. Prigogine (e.g., 1978) called such systems dissipative structures. The root of the idea goes back at least to the pre-Socratic Heraclitus (536 B.C.) who, in contrast to Parmenides, for whom true reality was entirely static, characterized the world as a continual process of transformational flow, and its objects as constituted by a generalized metabolism or combustion. Centuries later, in *De Anima*, Aristotle, stressing the active agency and generalized metabolism, consumption, growth, and decay of such systems, said of fire that it "alone of the primary elements [earth, water, air, and fire] is observed to feed and increase itself" (1947, p. 182).

Source: Photo courtesy of the National Severe Storms Laboratory.

Figure 2. A tornado is an example of an autocatakinetic system, a dynamically ordered flow structure whose identity, in contrast to a machine or artifact, is constituted not by a set of particular components typically occupying fixed positions with respect to each other, but by the ordered relations maintained by the incessant flow of its components. The dynamical order that defines the persistence of an autocatakinetic system as an object at the macro level is maintained through constant change at the micro level. This incessant flux of components can be thought of as a generalized metabolism by which the system maintains itself by pulling environmental potentials (or resources) into its autcatakinesis, which it returns in a more dissipated form. All living things from bacteria to human cultural systems, as well as the planetary system as a whole, which maintains a constant level of oxyen by this same generalized process, are members of the class of autocatakinetic systems.

In modern times, the idea was picked up by Leibniz who, following Heraclitus, described the dynamical persistences of the world as in a state of "perpetual flux, like rivers [where] the parts are continually entering in and passing out" (Rescher 1967, p. 121). The idea was first used as part of a general theory of evolution by Spencer (1852; Swenson In press-b).

Figure 3 shows a schematic of a generalized autocatakinetic system. Circular causality, as in closed-circle theory, and its various relatives

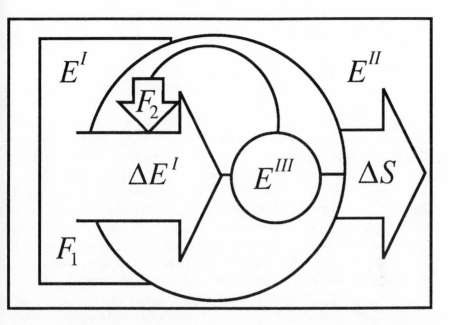

Source: Adapted from Swenson (1989b, p. 191). Copyright © 1989 by Pergamon. Adapted by permission.

Figure 3. A generalized autocatakinetic system. E^I and E^{II} indicate a source and a sink with the difference between them constituting a field potential with a thermodynamic force, F_1 (a gradient of a potential), the magnitude of which is a measure of the difference between them. ΔE^I is the energy flow at the input, the drain on the potential which is transformed into entropy production ΔS at the output. E^{III} is the internal potential carried in the circular relations that define the system by virtue of its distance from equilibrium that acts back to amplify or maintain input during growth or non-growth phases respectively with an internal force F_2.

play a central role in autocatakinetic systems, but in contrast to the autonomous circular relations of closed-circle theory which refer only to themselves, the circularity that defines an autocatakinetic system defines and maintains it in relation to its environmental sources. Autocatakinetic systems are embedded circles whose existence is inseparable from their environments both in actuality and by definition. In contrast to generalized Cartesian or closed circles, the circularity that defines the existence of autocatakinetic systems refers to the autocatakinetic-environment relation. There is no existence or self-reference for an autocatakinetic system independent of this relation. The rest of this section sketches some of the important generic behavior of autocatakinetic systems, and the following section describes the nomological basis for this dynamical ordering and the way it is manifested in the intentional dynamics of living things.

Order Production, Symmetry Breaking, and Space-Time Dimensions

Simply put, symmetry is invariance over change. Something is symmetric under certain operations or transformations if those operations leave it unchanged—if it remains the same or, put differently, is conserved under those operations. The greater the number of symmetry operations that can be performed on a thing to which it is indifferent or remains unchanged, the greater its symmetry. With geometric objects, for example, a sphere has greater symmetry than any other with respect to its rotational symmetry group because it is left invariant under arbitrary rotations around any axis passing through its center. Because these rotations can take on any value, the rotational symmetry group of a sphere is said to be continuous. In contrast, the symmetry group of a cube is discrete rather than continuous, and its symmetry is considerably lower. It is symmetric only under rotations around an axis through its face centers of 90 degrees, 180 degrees, 270 degrees, and 360 degrees (fourfold rotations). As this example shows, discontinuities constitute a break or reduction in symmetry, and from this we see that spontaneous order production—the appearance of an autocatakinetic system where there was none before—constitutes a symmetry-breaking event. There is now an object where there was no object before, and such an object constitutes a discontinuity in the field or environment from which it arises. When a tornado comes into being in a sky where there previously was no tornado, it breaks the symmetry of the sky.

This is further illustrated with a classic laboratory example of spontaneous ordering, or self-organization, known as the Bénard experiment (see Figure 4). In this experiment, a viscous fluid (silicone oil) is placed in a dish and heated uniformly from below. As a consequence of the difference in temperature, or gradient, between the hot bottom (source) and the cool air on top (sink), a potential exists which results in a flow of energy as heat from source to sink. Figure 4 shows two time slices from this experiment. The left-hand photo shows the disordered or Boltzmann regime where the potential is below a minimal threshold, and the source-sink flow is produced by the random, or disordered, collisions of molecules. In this regime, the surface of the system is smooth, homogeneous, and symmetrical. Any part can be exchanged with any other without changing the

Source: Swenson (1989b, p. 192). Copyright © 1989 by Pergamon. Reprinted by permission.

Figure 4. Two time slices from the Bénard experiment. The first time slice (left) shows the homogeneous, or disordered, "Boltzmann regime," where entropy is produced by heat flow from the disordered collisions of the molecules (by conduction), and the second (right) shows entropy production in the ordered regime. Spontaneous order arises when the field potential is above a minimum critical threshold, and stochastic microscopic fluctuations are amplified to macroscopic levels as hundreds of millions of molecules begin moving in an orderly fashion together.

appearance or dynamics of the system at all. When the potential is increased beyond the critical threshold, however, the situation changes dramatically as spontaneous order arises and the symmetry of the disordered regime is broken. The dynamical ordering of the system produces macroscopic discontinuities with distinct space-time orientations that make it no longer possible to arbitrarily exchange one part for another.

The relation between order production, symmetry breaking, and space-time dimensions is an important one and can be brought further into focus by looking at the Bénard experiment in more detail. Figure 5 shows the ordered autocatakinetic flow of molecules constituting an individual Bénard cell. Here, by way of the stream lines, we can see in detail the way the continuous flow of components at the microscopic level constitutes the structure at the macroscopic level. As this figure helps visualize, because the intrinsic space-time dimensions for any system or process are defined by the persistence of its component relations, the transformation from disorder to order increases its dimensions dramatically. Put in different terms, the symmetry breaking that occurs in the production of order from disorder implies a dramatic increase in a system's space-time dimensions.

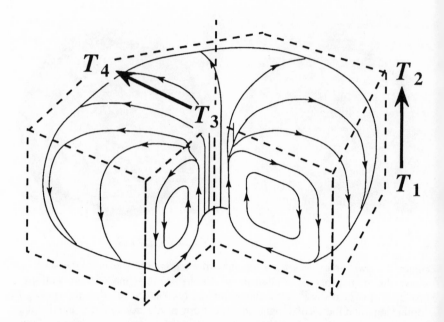

Figure 5.   The autocatakinetic flow of the fluid constituting a Bénard cell is shown by the small arrows. $T_1 \rightarrow T_2$ is the heat gradient between the heat source below and the sink above that constitutes the potential that motivates the flow. Because density varies inversely with temperature, there is also a density gradient from bottom to top giving groups of molecules ("parcels") that are displaced upwards by stochastic collisions an upward buoyant force. If the potential is above the minimum threshold, parcels will move upward at a faster rate than their excess heat can be dissipated to their surrounds. At the same time, such an upward flow of heat will increase the temperature of the upper surface directly above it, creating a surface tension gradient $T_3 \rightarrow T_4$ which will act to further amplify the upward flow by pulling the hotter fluid to the cooler surrounds. The upward displacement of fluid creates a vacuum effect pulling more heated fluid from the bottom in behind it, which in turn makes room for the fluid which has been cooled by its movement across the top, to fall, be heated, and carry the cycle on, and autocatakinesis has been established.

   In the ordered regime of the Bénard experiment, the intrinsic space-time dimensions are of the order of seconds and centimeters. It takes the fluid some seconds to make an autocatakinetic cycle between source and sink, and the distance covered, or the dimensions of a single cell, such as that shown in Figure 5, can be measured in centimeters. This is in stark contrast to the disordered regime where the intrinsic space-time dimensions are defined by mean free-path distances and relaxation times (the distances and times between random or disordered collisions) and are on the order of 10^{-8} centimeters and 10^{-15} seconds. From this, it is seen that with the breaking of symmetry in the production of spontaneous order, the system accesses and fills new dimensions of space-time beyond the reach of its previous regime. The same generic dynamics can be seen with respect to terrestrial evolution in Figure 1. Here, spontaneous ordering occurs at symmetry-breaking events as minimal critical thresholds of atmospheric oxygen are reached with the system, as a consequence progressively filling new dimensions of space-time and moving, contrary to the Boltzmann interpretation of the second law, increasingly further from thermodynamic equilibrium. This relationship between spontaneous ordering and the filling or extension of space-time dimensions, as the final section of this paper will show, provides an important piece to the apparent puzzle of the river that flows uphill. From this, evolution on Earth can be seen as a process of symmetry-breaking events by which the terrestrial system as a whole accesses new dimensions of space-time and moves progressively further from equilibrium. This provides a set of observables that establishes the direction or time-asymmetry of evolution.

Insensitivity to Initial Conditions, Downward Causation, or Macrodeterminacy, and the Genericity of Populations of One

The preceding sections dealing with closed-circle theory and Darwinism as the theory of evolution showed a number of major problems or anomalies that render both of these approaches inimical to a comprehensive evolutionary theory, to an account of intentional dynamics, or to the active epistemic dimension of the world. As a consequence, the approaches are inimical to ecological theory. Given the anti-ecological Cartesian postulates at each of their cores, this is inevitable. Neither proposes a universal embedding, by which I mean an embedding in a physical world that is commensurable with the behavior the approaches would like to explicate. By contrast, the study of autocatakinetic systems implies commensurability. By their definition and by their behavior, they exist through, and as differentiations of, the larger systems or world from which they arise.

Living systems are a kind of autocatakinetic system. In particular, they are autocatakinetic systems with replicating components. Autocatakinesis, self-organization, or spontaneous ordering, however, is a universal property that is not dependent on, and therefore is not explained by, replicating components—or, in different terms, is not explained by biology or culture.

It is the universality of spontaneous ordering, or autocatakinesis, that provides the basis for understanding the commensurability of all self-organizing systems in general. In this subsection, although it is understood that the fact of autocatakinetic systems (viz., the nomological basis for the river that flows uphill in relation to the river that flows down) remains to be explained until the next section, it will be shown here that the major problems or anomalies of insensitivity to initial conditions, downward causation, and populations of one, are generic properties and behaviors— everyday expected behavior—and are not anomalies or problems within the context of autocatakinetic systems.

Real-world systems—particularly, but not by any means exclusively, living things and the intentional dynamics that distinguish them—are remarkably insensitive to initial conditions. Because orthodox theory adheres to an impoverished causal description of the world—namely, that it is essentially microdetermined—it has no basis to admit what amounts to macroscopic causality or downward causality into its explanatory framework. It is for this same reason that it cannot address the problem of the population of one. Put in simple and blunt terms, it fails to recognize the universality of autocatakinesis, or self-organization, and assumes with its Cartesian postulates and Boltzmannian thermodynamics an incommensurable physics. Insensitivity to initial conditions, downward causality, and macrodeterminism are generic properties of autocatakinetic systems. We return to the Bénard experiment again, in more detail, for an illustration.

Returning to Figure 4, the right-hand photo shows the system filled with Bénard cells of variable size and shape shortly after the critical threshold has been crossed. As time continues, however, a spontaneous process of selection occurs that includes the subsumption of smaller cells by larger ones, the competitive exclusion of smaller cells by larger ones, and the spontaneous division, or fission, of larger cells to smaller ones (e.g., see Swenson 1989a, 1989b, 1992, In press-c, for the time-series). The end result is a regular array of hexagonal cells of uniform size and shape. Now, the point to make is that the variability that is seen in Figure 4, which is at the beginning of the process, is a consequence of the fact that order production is stochastically, or randomly, seeded. The end state, however, is macrodetermined.

In particular, in the disordered regime the dynamics are characterized by random collisions between microcomponents which constitute fluctuations around an average state. When the critical threshold is crossed spontaneous order is seeded by any fluctuation anywhere in the fluid that is of a minimal amplitude. Since the location and actual amplitude of such fluctuations are stochastically determined, the cells will form at different places in the fluid and will grow at different rates every time the experiment is done. Seconds after the critical threshold is crossed, the fluid thus fills with cells of variable size, but each and every time the experiment is run, the variability in the size and shape is progressively eliminated by a process of selection to produce a final state of regularly arrayed hexagonal cells of uniform size and shape. In a decidedly non-Laplacian fashion, dissimilar micro-antecedents lead to similar macroscopic consequences. Here we see a process of "blind variation" in the stochasticity of the microcomponents in the disordered regime, and a lawful process of selection leading to a macrodeterminate result. Random initial conditions at the micro level do not mean that the evolution of the system is random or undetermined. Initial conditions, which can vary dramatically relative to their own frame of reference, need only meet some minimal general conditions, and the laws of form do the rest.

A number of other generic properties that can be observed in this example bear pointing out. When the critical threshold is reached in the Bénard cell experiment and the fluid fills with cells, every cell arises initially as a population of one. The population of one is not anomalous with respect to autocatakinesis: Autocatakinetic systems *are* populations of one, and the general conditions for the establishment of autocatakinetic systems are generic across scales. In each case, this involves: (1) stochasticity or "blind variation" at the micro level that "seeds" order at the macro level; (2) circular causality that amplifies the microscopic seeding to establish autocatakinesis at the new macroscopic level; and (3) a source-sink gradient above some minimal critical level sufficient to pump up or fill out the new dimensions of space-time that the establishment and maintenance of autocatakinesis entails. The specific details of the establishment of macroscopic order in the Bénard experiment are discussed in the legend to Figure 5.

As the generic description implies, autocatakinetic systems are deviation-amplifying systems, to use Maruyama's (1963) term. They come into being as a consequence of positive feedback which acts to amplify small deviations or displacements away from thermodynamic equilibrium. Although negative feedback and homeostasis follow naturally from positive feedback as a consequence of various limits to growth or laws of form that follow from the finite nature of space-time, autocatakinetic systems come into being and are characterized by growth and by the departure from thermodynamic equilibrium. Typically, providing sufficient environmental potential exists when the system reaches a limit (a critical minimal threshold), order production continues either horizontally, by fissioning, or vertically through the production of a new macroscopic level.

Because autocatakinetic systems are dependent on their surfaces for pulling in environmental potential, and because in isometric growth surfaces increase as the square of a linear dimension while the volume increases as the cube, some form of surface-volume law, or related laws of form, typically determines a minimum and a maximum size that a system can be before fissioning. Again, fluctuations play an important role in the symmetry-breaking process. Below a critical threshold, they are dampened; and above it, they are amplified. This generic order-producing dynamic is seen from simple physical systems, such as the Bénard experiment (see Swenson 1989a, 1989c, 1992, for photos of fissioning of Bénard cells) to bacteria, and through to the autocatakinesis of cultural ordering and planetary autocatakinesis as a whole.

From early Paleolithic to early Neolithic times, to take a cultural example, the hominid population increased from some few tens of thousands to something like 5-10 million, but not through a corresponding increase in the size of autonomous communities (not by building new levels of order, or vertical ordering or growth). Rather, it was through the proliferation by fissioning of the number of communities (i.e., by horizontal growth), from something like 1,500 at the beginning of the Paleolithic to some 75,000 or so at the end (Carneiro 1987). The fissioning of autonomous villages, given a supply of initial conditions within tolerance limits, as with the Bénard case, is a macrodeterminate process. Below a critical size or threshold, social interactions which can be thought of as fluctuations or deviations from the mean (e.g., adultery, theft, disharmonious acts of witchcraft) are damped. When an autonomous unit exceeds a certain minimal size, however, these same microconditions are amplified to macroscopic proportions and fissioning occurs. This fissioning was the almost exclusive means of growth of human culture for some 99 percent of its history until suddenly, and within a short period of time, after certain critical environmental thresholds were reached, vertical ordering occurred when previously autonomous units were pulled into the emergence of nation states, not once but repeatedly and independently, in numerous separate locations. As Carneiro has shown "[w]here the appropriate conditions existed, the state emerged...[and, for example, in the Valley of Mexico, Mesopotamia, the Nile Valley, and the Indus Valley] the process occurred in much the same way for essentially the same reasons" (1970, p. 733).

Spontaneous Ordering Occurs Whenever It Gets the Chance

Finally, let us return to the Bénard experiment to emphasize perhaps the most important point with respect to spontaneous order production. Here, it can be seen that order arises, not infinitely improbably, but with a probability of one, which is to say it arises every time, and as soon as the critical threshold is reached. Spontaneous ordering occurs, in other words, as soon as the opportunity arises. This conforms with the biological extremum (the fecundity principle) that takes the production of as much biological order as possible to be the "inherent property" of life, and the evolutionary record writ large. It suggests that the production of higher-ordered forms, including the origin of life itself occurred, not as a repeated series of astronomically improbable accidents (which certainly would be "infinitely improbable"), but as soon as it had the chance—that the origin of life on Earth appeared not after some long lifeless time but as soon as the Earth was cool enough to support oceans, and that the higher-ordered forms appeared as soon as minimal levels of atmospheric oxygen were reached (Figure 1).

If the world in general produces as much order as it can, what is the nomological basis? The answer is given in the next section, and it provides the principled basis for unifying the two otherwise apparently incommensurable rivers.

WHY THE WORLD IS IN THE ORDER-PRODUCTION BUSINESS: THE NOMOLOGICAL BASIS FOR INTENTIONAL DYNAMICS

Symmetry and Broken Symmetry Again: The Classical Statements of the First and Second Laws of Thermodynamics

The insistence of Heraclitus on the importance of persistence and change, and his view of the world as an ongoing process of flow, would certainly seem to qualify him as the foremost ancient progenitor of what would become the science of thermodynamics. Leibniz's assertion that there must be something that changes and something that remains the same, or something conserved, and his very aggressive work to identify the conserved and active quantities of the world, certainly qualify him as the modern founder of thermodynamics. The work of Mayer and also Helmoholtz, who among others are credited with formulating the first law, can be traced in a direct lineage to Leibniz. One may also locate legitimate roots for the laws of thermodynamics in those who searched for symmetry principles, Parmenides among them, because the first and second laws, understood in the deepest sense, are symmetry principles. Eddington (1928) has argued that the second law holds the supreme position among all the laws of nature, but it is probably more accurate to say that the first and second laws together hold the supreme positions among all the laws of nature, because they are each dependent in a certain way upon the other.

Following the earlier work of Davy and Rumford, the first law was first formulated by Mayer, then Joule, and later Helmoholtz in the first half of the nineteenth century with various demonstrations of the equivalence of heat and other forms of energy. The law was completed in this century with Einstein's demonstration that matter is also a form of energy. With its recognition that all natural processes can be understood as flows or transformations of different forms of energy, and that the total quantity of energy always remains the same, or is conserved, the first law provided the basis for unifying all natural processes through the recognition of their underlying time-translation symmetry. The first law, in other words, expresses what remains the same through all natural processes, regardless which way one goes in time. This presumably would have made Parmenides happy because as far as the first law goes, nothing changes—or, in other words, there is no time. When the potential energy of an elevated body of water is, by its fall, turned into mechanical energy to drive a mill wheel, and the mechanical energy in turn is dissipated into the surrounds as heat from the friction of the millstone, the total amount of energy is conserved, or has remained unchanged, and that is what the first law says (that energy is never created or destroyed is, thus, another statement of the first law).

Until Clausius and Thomson (who later became Lord Kelvin) came along, there was nevertheless some confusion and doubt about this law. This was because, as Joule's experiment (see Figure 6) demonstrating the conservation of energy unintentionally showed, there is a broken symmetry to natural processes, a one-way flow of things that, in contrast to the first law, establishes the notion of time, or a difference between past, present, and future. The same is easily seen with the example of the mill wheel. It was the relation between the symmetry on the one hand, and broken symmetry on the other, that Clausius and Thomson showed with their formulation of the second law in the 1850s. The work of Carnot, some 25 years earlier, brought the problem to a head. Carnot had observed that, like the fall of a stream that turns a mill wheel, it was the "fall" of heat from

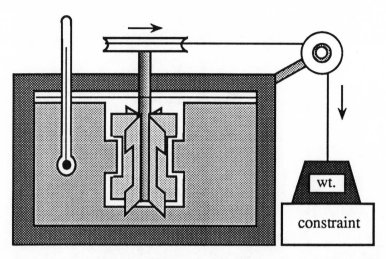

Figure 6. Experiment devised by Joule to show the conservation of energy. When a constraint is removed, potential energy in the form of a suspended weight is converted into the mechanical or kinetic energy of a moving paddle wheel in an energy-tight container of water, heating the water by an amount consistent with the amount of potential energy lost by the falling weight.

higher to lower temperatures that motivated a steam engine. That this work showed an irreversible destruction of "motive force," or potential for producing change, suggested to Clausius and Thomson that if the first law was true, then contrary to popular misconception energy could not be the motive force for change. Recognizing in this way that the active principle and the conserved quantity could not be the same, they realized that there must be a second law involved. Clausius coined the word entropy to refer to the dissipated potential, and the second law states that all natural processes proceed so as to maximize the entropy (or, equivalently, minimize or dissipate the potential), while at the same time energy is entirely conserved. The balance equation of the second law, expressed as

$$\Delta S > 0$$

says that in all real-world processes, entropy always increases.[2,3]

The active nature of the second law is intuitively easy to grasp and empirically easy to demonstrate. Figure 7 shows a glass of hot liquid placed in a room at a cooler temperature. The difference in temperatures in the glass-room system constitutes a potential, and a flow of energy in the form of heat—a "drain" on the potential—is produced from the glass (source) to the room (sink) until the potential is minimized (the entropy

$$T^{I} > T^{II}$$

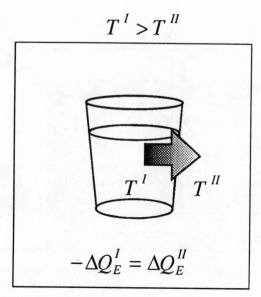

$$-\Delta Q_{E}^{I} = \Delta Q_{E}^{II}$$

Source: Swenson (1991, p. 45). Copyright © 1991 Intersystems Publications. Adapted by permission.

Figure 7.   A glass of liquid at temperature T^I is placed in a room at temperature T^{II} such that $T^I > T^{II}$. The disequilibrium produces a field potential that results in a flow of energy in the form of heat $-\Delta Q_E^I$ from the glass to the room so as to drain the potential until it is minimized (the entropy is maximized), at which time thermodynamic equilibrium is reached and all flows stop. $-\Delta Q_E^I = \Delta Q_E^{II}$ refers to the conservation of energy in that the flow from the glass equals the flow of heat into the room.

maximized) and the liquid and the room are at the same temperature. At this point, all flows and, thus, all entropy production stop, and the system is at thermodynamic equilibrium. The same principle applies to any system where any form of energy is out of equilibrium with its surrounds (e.g., mechanical, chemical, electrical, or energy in the form of heat), a potential that real-world processes act spontaneously to minimize.

The Second Law as a Law of Disorder

The active macroscopic nature of the second law presented a direct challenge to the "dead" mechanical world view. Boltzmann tried to meet the challenge by reducing the law to a statement of probability following upon the random collisions of mechanical particles. Following Maxwell, and modeling gas molecules as colliding billiard balls in a box, Boltzmann noted that with each collision, nonequilibrium velocity distributions (groups of molecules moving at the same speed and in the same direction) would become increasingly disordered, leading to a final state of macroscopic uniformity and maximum microscopic disorder: the state of maximum entropy (where the macroscopic uniformity corresponds to the obliteration of all field potentials). The second law, Boltzmann argued, was thus simply the result of the fact that in a world of mechanically colliding particles, disordered states are the most probable. Because there are so many more possible disordered states than ordered ones, a system will almost always be found either in the state of maximum disorder—the macrostate with the greatest number of accessible microstates, such as a gas in a box at equilibrium—or moving toward it. A dynamically ordered state, one with molecules moving "at the same speed and in the same direction," Boltzmann concluded, is thus "the most improbable case conceivable…an infinitely improbable configuration of energy" (1974 [1886], p. 20).

Boltzmann himself acknowledged that his hypothesis of the second law had only been demonstrated for the case of a gas in a box near equilibrium, but the science of his time (and up until quite recently) was dominated by linear, near-equilibrium, or equilibrium thinking, and this view of the second law, as a law of disorder, became widely accepted. The world, however, is not a linear, near-equilibrium system like a gas in a box, but instead is nonlinear and far from equilibrium, and the second law is not reducible to a stochastic collision function. As the next subsection outlines, rather than being infinitely improbable, we now can see that spontaneous ordering is the expected consequence of natural law.

Why the World Is in the Order-Production Business

The idea that living things violate the second law of thermodynamics was temporarily deflected in the middle of this century when Bertalanffy showed that "spontaneous order…can appear in [open] systems" (1952, p. 145)—that is, systems with energy flows running through them—by virtue of their ability to build their order by dissipating potentials in their environments. As briefly noted above, along the same lines, pointing to the balance equation of the second law, Schröedinger (1945) popularized the idea of living things as streams of order which like flames are permitted to exist away from equilibrium because they feed on "negentropy" (potentials) in their environments. These ideas were further popularized by Prigogine (e.g., 1978).

Schrödinger's important point was that as long as living things like flames (and all autocatakinetic systems) produce entropy (or minimize potentials) at a sufficient rate to compensate for their own internal ordering or entropy reduction (their ordered departure and persistence away from equilibrium), then the balance equation of the second law, which simply says that entropy must increase in all natural processes, would not be violated. According to the Bertalanffy-Schröedinger-Prigogine view, order *can* arise spontaneously, and living things are thus *permitted* to exist, as it became popular to say, so long as they "pay their entropy debt." While this made an important contribution to the discourse and worked for the classical statement of the second law per Clausius and Thomson, in Boltzmann's view such "debt payers" were still infinitely improbable. Living things were still infinitely improbable states struggling or fighting against the laws of physics. The urgency toward existence captured in the fecundity principle and the intentional dynamics it entails, as well as planetary evolution as a whole, were still entirely anomalous on this view with respect to universal law. What is more, as the Bénard experiment shows, simple physical systems also falsify the Boltzmann hypothesis. Order is seen to arise, not infinitely improbably, but with a probability of one, that is, whenever, and as soon as it gets the chance. The nomological basis for this opportunistic ordering was still a mystery, a point emphasized by Bertalanffy himself, who suspected there might be another thermodynamic principle that would account for this "build-upism" (Koestler 1969, p. 52) or, in the terms we have been using, the river that flows uphill.

Space-Time Relations, Order Production, and a Return to the Balance Equation of the Second Law

There are two key pieces to solving the puzzle or problem of the two incommensurable rivers. The first is discovered by returning to the balance equation of the second law. As discussed above and illustrated in Figure 5, transformations from disorder to order dramatically increase the space-time dimensions of a system. What Bertalanffy and Schröedinger emphasized was that as long as an autocatakinetic system produces entropy fast enough to compensate for its development and maintenance away from equilibrium (its own internal entropy reduction or increase in space-time dimensions), it is permitted to exist. Ordered flow, in other words, to come into being or exist must function to increase the rate of entropy production of the system plus environment at a sufficient rate—it must pull in sufficient resources and dissipate them—to satisfy the balance equation of the second law. This implicitly makes an important point, which was not spe-

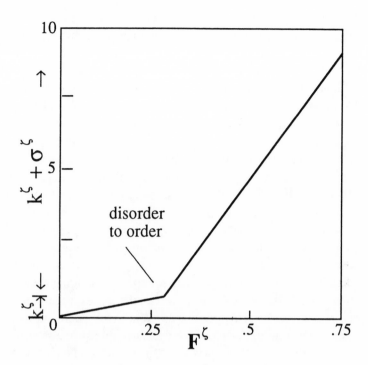

Figure 8. The discontinuous increase in the rate of heat transport that follows from the disorder-to-order transition in a simple fuild experiment similar to that shown in Figure 4. The rate of heat transport in the disordered regime is given by k^ζ, and $k^\zeta + \sigma$ is the heat transport in the ordered regime [$3.1 \times 10^{-4}H$ (cal × cm.$^{-2}$ × sec^{-1})].

cifically noted by Bertalanffy or Schroedinger and which can now be stated explicitly: *Ordered flow must be more efficient at dissipating potentials than disordered flow.* Figure 5 shows exactly how this works in a simple physical system. Figure 8 shows the dramatic increase in the rate of heat transport from source to sink that occurs in the transformation from the disordered to ordered state. Given the balance equation of the second law, the superior dissipative efficiency of ordered flow could not be otherwise. This important point brings us to the second and final piece of the puzzle.

The Law of Maximum Entropy Production

The crucial final piece to the puzzle of the two rivers, the one that provides the nomological basis for dissolving the postulates of incommensurability, is the answer to a question that classical thermodynamics never asked. The classical statement of the second law says that entropy will be maximized, or potentials minimized, but it does not ask or answer the question of which of the available paths a system will take to accomplish this end. The answer to the question is that *the system will select the path, or assembly of paths out of otherwise available paths, that minimizes the potential or maximizes the entropy at the fastest rate given the constraints.* This is a statement of the law of maximum entropy production and is the physical principle that provides the nomological basis, as will be seen below, for why the world is in the order-production business (Swenson 1988, 1989d, 1991a, 1991b, 1992, 1996; Swenson and Turvey 1991). Note that the law of maximum entropy production is in *addition* to the second law. The second law says only that entropy is maximized, while the law of maximum entropy production says it is maximized (potentials minimized) at the fastest rate given the constraints. These are two separate laws because the second, in principle, could be falsified without changing the first. Like the active nature of the second law, however, the law of maximum entropy production is intuitively easy to grasp and empirically demonstrate.

Consider the example of the warm mountain cabin sitting in cold, snow-covered woods (Swenson and Turvey 1991). The difference in temperature between the cabin and the woods constitutes a potential, and as a consequence, the cabin/woods system will produce flows of energy as heat from the cabin to the woods so as to minimize the potential. Initially, supposing the cabin is tight, the heat will be flowing to the outside primarily by conduction through the walls. Now imagine opening a window or a door and thus removing a constraint on the rate of dissipation. What we know intuitively, and can confirm by experiment, is that whenever a constraint is removed and a new path or drain is provided that increases the rate at which the potential is minimized, the system will seize the opportunity. Furthermore, since the opened window will not instantaneously drain all the potential, some heat will still be allocated to conduction through the walls. Each path will drain all that it can, the fastest (in this case, the open window) procuring the greatest amount with the remainder going to the slower paths (in this case, conduction through the walls). In other words, regardless of the specific conditions or the number of paths or drains, the system will automatically select the assembly of paths from among those otherwise available so as to get the system to the final state, to minimize or drain the potential, at the fastest rate given the constraints. This is the essence of the law of maximum entropy production. Now, what does this have to do with spontaneous ordering, with the filling of dimensions of space-time?

Given the preceding, the reader may have already leaped to the correct conclusion. *If* the world selects those dynamics that minimize potentials at the fastest rate given the constraints (the law of maximum entropy production), *and if* ordered flow is more efficient at reducing potentials than disordered flow (derivation from the balance equation of the second law), *then* the world can be expected to produce order whenever it gets the chance. *The world is in the order-production business because ordered flow produces entropy faster than disordered flow* (Swenson 1988, 1991, 1992, 1996; Swenson and Turvey 1991). Contrary to the older Boltzmann view where the production of order is seen as infinitely improbable, given this new understanding, the world can be expected to produce as much order as it can, which is to say, to expand space-time dimensions whenever the opportunity arises. Autocatakinetic systems, in other words, are self-amplifying sinks that pull potentials or resources into their own development and persist away from equilibrium by extending the space-time dimensions of the fields (system plus environment) from which they emerge, and thereby increase the dissipative rate. The law of maximum entropy production, when coupled with the balance equation of the second law, provides the nomological basis for dissolving the postulates of incommensurability, and unifying living things with their environments—for unifying the two otherwise apparently incommensurable rivers that flow up and downhill, respectively. Rather than an incommensurable, inexplicable, and infinitely improbable anomaly—the river that flows uphill—the active ordering that characterizes terrestrial evolution, of which biological and cultural evolution are components, is seen to be an expected manifestation of universal law.

End-directed Behavior Dependent on Meaning

There is an extremely important property of the intentional dynamics of living things, or of the river that flows uphill, that remains to be addressed. At the beginning of this paper, intentional dynamics were defined as end-directed behavior prospectively controlled or determined by meaning, or information about paths to ends, and this was contrasted with end-directed behavior which can be understood as determined by local potentials and fundamental laws. Examples of the latter were a river flowing down a slope or heat flowing down a gradient. We can elaborate this discussion, given what we have covered in the interceding pages, by including examples of autocatakinetic systems such as the Bénard experiment, tornadoes, and dust devils, systems that we call self-organizing but do not say are characterized by intentional dynamics. The autocatakinesis of such systems, which breaks symmetry with previously disordered regimes to access and dynamically fill higher-ordered dimensions of space-time, is still determined with respect to local potentials with which they typically remain permanently connected. The autocatakinesis of living things, in contrast, is maintained with respect to non-local potentials, potentials discontinuously located in space-time to which they are not permanently connected (Swenson 1991b, In press-a; Swenson and Turvey 1991).

If we understand from universal principles that the world acts, in effect, to maximize its extension into space-time, or to produce as much order as possible, we can see immediately what intentional dynamics provide. By providing the means for linking together or accessing and dissipating discontinuously located, or non-local, potentials in the building of order, intentional dynamics provide access to vast regions of space-time that are otherwise inaccessible. Just as there is a qualitative leap in the transformation of disorder to order, with respect to the potentially accessible dimensions of space-time it offers, intentional dynamics constitute a symmetry-breaking or qualitative leap in terrestrial order production. Likewise, the origin of human cultural systems which with highly developed symbolic langauge that may be thought of as intentional dynamics about intentional dynamics, provides dramatic access to new dimensions of space-time— also a terrestrial symmetry-breaking event (see Dyke's [1997, this volume] discussion on the increase of space-time dimensions in human cultural systems).

In the section on evolution, it was shown that the assertion of evolutionary epistemologists that evolution constitutes a progressive knowledge-acquisition process from amoeba to Einstein, was an assertion that could not be made (nor accounted for) on the grounds of Darwinian theory. According to Darwinian theory, amoebae and Einsteins are incommensurable and hence, like Kuhnian paradigms or generic closed-circles, are incomparable. It was pointed out that evolution would have to be about something other than fitness to make the assertion that evolutionary epistemologists would like to make. Our understanding that from a universal standpoint, terrestrial evolution is a planetary process about entropy production maximization, and as a consequence the filling of space-time dimensions, provides the principled basis to make the assertion. Terrestrial evolution is indeed a progressive knowledge-acquisition process from amoeba to Einstein (more appropriately, from Archean prokaryotes to the contemporary globalization of human culture) through which the system learns, in effect, to access new, otherwise inaccessible, space-time dimensions.

But now, the part that still needs explaining: If intentional dynamics are not determined by local potentials, then how are they determined? To simply say they are meaningfully determined, at this point, begs the question. Autocatakinetics has the property of insensitivity to initial conditions, or the property of macrodeterminacy, but what is the basis for the macrodeterminacy of intentional systems if not local potentials? The Bénard convection, which in effect "solves the packing problem" by producing a regular array of hexagonal cells during the course of its evolution or development, can be understood in terms of the system's proximal relation to, or embeddedness within, a field of local potentials. But, how is intentional behavior determined with respect to non-local or distal potentials? How does it solve the packing problem with respect to non-local potentials? What is the physical basis for the epistemic relations by which the accessibility of new space-time levels of order are effectively opened up? How, in other words, does one get from an otherwise meaningless world of extension, or usual physical description, to a meaningful world of intension?

From Extension to Intension

We return to our first principles—in particular, first-law symmetry, second-law broken symmetry, and the law of maximum entropy production as ordering principle—for immediate clues. First, we recognize that, consistent with thermodynamic inquiry, the search here is for macroscopic observables. Autocatakinetic systems are macroscopic systems, embedded in macroscopic flow, and the search is thus not for "meaning" in individual particles but for macroscopic flow variables that capture invariant properties with relevance to intentional ends. Following the same methodology suggests, further, that the search for macroscopic observables involves a search for symmetry and broken symmetry—for observables that capture the nomological relation between persistence and change of the distal objects of intention with respect to the proximal or local space-time position of the epistemic subject. It turns out this is exactly the insight of Gibson's (1986; Swenson and Turvey 1991; Turvey and Shaw 1995) ecological conception of information. The idea developed by Gibson with respect to animals and their environments has now been extended to life in general and embedded in a universal thermodynamic context by "neo-Gibsonians" and "third-wave Gibsonians" (e.g., Peck In press; Swenson In press-a; Swenson and Turvey 1991; Turvey and Shaw 1995). The core idea is deceivingly simple but has profound explanatory consequences.

Living things are embedded in ambient energy flows (e.g., optical, mechanical, chemical) for which the mean energy content is extremely low relative to the energy used by living things from their on-board potentials to power their intentional acts. As a consequence of first-law symmetry, lawful or invariant relations exist between the macroscopic properties of such ambient energy distributions and their sources, with the further consequence that the former can be used in the prospective control of intentional ends to specify or determine the latter. A chemical gradient that lawfully specifies the source of their food can be used by bacteria, diffusion fields of diffusing volatiles that lawfully specify the sources of their intentional ends may be used by animals, and fields of mechanical waves and optical fields can be used in similar ways.

A particularly crucial and widespread requirement for the intentional dynamics of many living things is the ability to effect controlled collisions. Examples include soft collisions with little or no momentum exchange, as in a bird landing on a branch; hard collisions with substantial momentum exchange, as when a predator attacks a prey; and collision avoidance, where the ends of an intentional agent require that it not collide with particular things. The fact of first-law symmetry means that "information about" such collisions is lawfully carried in the ambient energy field (the "optical flow field") that transforms itself as a living thing moves through it. Just as in the Bénard case, where local potentials and laws of form specify the origin, production, and development of order, so too it is with nonlocal potentials and the invariant or epistemic properties of ambient energy flows with respect to intentional dynamics.

Following the case of controlled collisions further, the time-to-contact (τ) as shown in Figure 9 is determined by the inverse of the relative rate of expansion of the optical flow field, and the information about whether a collision will be hard or soft is given by the time derivative or rate of change of the relative rate of expansion ($\dot{\tau}$) (Lee 1980; Kim et al. 1993). In the case of a bird landing on a branch and requiring a soft collision, for example, the rate of change must be

$$(\dot{\tau}) \geq -.5.$$

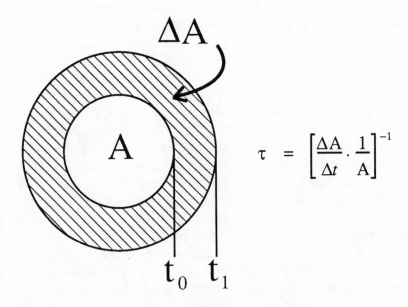

$$\tau = \left[\frac{\Delta A}{\Delta t} \cdot \frac{1}{A}\right]^{-1}$$

Figure 9. Time-to-contact, τ, is determined by the inverse of the relative rate of expansion of the optical flow field, A.

This example shows how a single macroscopic variable nomologically carried in the optic flow can precisely determine the intentional dynamics of living things—in this example, when a particular bird must open its wings to decelerate now so that it does not, in effect, crash into a branch later. This deceptively simple understanding exposes the fact that not only are the shapes and forms things assume nomologically determined by laws of form (e.g., there is, within tolerance, a requisite ratio between flight muscle weight and body weight, or between wing span and body weight, or between brain weight and body weight [e.g., Alexander 1971]) but that information about, or meaning, carried in macroscopic flow variables, nomologically determines the behavior of things toward their intentional ends.

CONCLUSION

Ecological science addresses the relations of living things to their environments, and the study of human ecology addresses the particular case of humans. There is an opposing tradition built into the foundations of modern science of separating living things and, in particular, humans from their environments. Beginning with Descartes' dualistic world view, this tradition found its way into biology by way of Kant, and into evolutionary theory through Darwin, and it manifests itself in two main postulates of incommensurability: the incommensurability between psychology and physics (the "first postulate of incommensurability") and that between biology and physics (the "second postulate of incommensurability").

The idea of the incommensurability between living things and their environments gained what seemed strong scientific backing with Boltzmann's view of the second law of thermodynamics as a law of disorder, according to which the transformation of disorder to order was said to be infinitely improbable. If this were true, and until very recently it was taken to be so, then the whole of life and its evolution becomes one improbable event after another. The laws of physics, in this view, predict a world that should be becoming more disordered, while terrestrial evolution is characterized by active order production. The world, in this view, seemed to consist of two incommensurable or opposing "rivers," the river of physics which flowed down to disorder, and the river of biology, psychology, and culture, which "flowed up," seemingly working to produce as much order as possible.

As a consequence of Boltzmann's view of the second law, evolutionary theorists up to present times have held onto the belief that "organic evolution was a negation of physical evolution" (Levins and Lewontin 1985, p. 69) and that biology and culture work somehow to "defy" the laws of physics (Dennett 1995). With its definition of evolution as an exclusively biological process, Darwinism separates both biology and culture from their universal, or ecological, contexts and advertises the Cartesian postulates of incommensurability at its core. These postulates are inimical to the idea of ecological science. An ecological science, by definition, assumes contextualization or embeddedness, and its first line of business must be to understand the nature of embeddedness. This requires a universal or general theory of evolution that can uncover and explicate the relationship between the two otherwise incommensurable rivers, and put the active ordering of biological and cultural systems, of terrestrial evolution as a time-asymmetric process back into the world.

The law of maximum entropy production, when coupled with the balance equation of the second law and the general facts of autocatakinetics, provides the nomological basis for such a theory. Together, they show why, rather than living in a world where order production is infinitely improbable, we live in and are products of a world that can be expected to produce as much order as it can. Together they show how the two otherwise incommensurable rivers—physics on the one hand, and biology, psychology, and culture on the other—are part of the same universal process. They show how the fecundity principle and the intentional dynamics it entails are special cases of an active, end-directed world opportunistically filling dynamical dimensions of space-time as a consequence of universal law. The epistemic dimension—the urgency toward existence, in Leibniz's terms— that characterizes the intentional dynamics of living things and is expressed in the fecundity principle, and the process of evolution writ large as a single planetary process, are thus not only commensurable with first, or universal, principles, are a direct manifestation of them.

The view presented here thus provides a principled basis for putting living things, including humans, back in the world and recognizing living things and their environments as single irreducible systems. It provides the basis for contextualizing the deep and difficult questions concerning the place of humans, as both productions and producers of an active and dynamic process of terrestrial evolution, which as a consequence of the present globalization of culture is changing the face of the planet at a rate which seems to be without precedent over geological time. Of course, answers to questions such as these always lead to more questions, but such is the nature of the epistemic process we call life.

ACKNOWLEDGMENTS

Special thanks as always to the Center for the Ecological Study of Perception and Action, and especially to Claudia Carello, Michael Turvey, and Bob Shaw for their work in providing an extraordinary environment conducive to the development of these ideas. Special thanks also to Lee Freese, Series Editor, and also Chuck Dyke for their perseverance in seeing that this article was brought to fruition. Preparation of this manuscript was supported in part by National Science Foundation Grand #SBR9422650. All errors are entirely mine.

NOTES

1. In fact, Descartes (1644/1975), recognizing the necessity of a conservation principle for a law-based physical world, proposed the conservation of "motion," which he thought would still allow him to get around the problem of interactionism. He thought "mind" could interact with "matter" by changing its direction but not the quantity of motion. Motion, as Leibniz (1696/1925), pointed out, however, is not a conserved quantity. It is momentum which is conserved, and momentum is a vector the conservation of which, like the conservation of energy, would be violated by exogenous interaction.

2.   It was Tait who first pointed out how counterintuitive it was to refer to the dissipative potential of a system as a quantity that increased, and he proposed reversing the sign so it would be possible to talk about entropy (as the potential for change) thus being minimized. Maxwell picked up on this, but it never caught on. Because the idea of entropy increase is oftentimes hard to conceive, in this text I will often use "minimize the potential" in addition to or instead of "maximize the entropy." They should be taken as equivalent expressions.

3.   Since its coinage by Clausius to refer to the dissipated potential in a system the word "entropy" has taken on numerous, and non-equivalent meanings. It is often used to refer to non-physical, as well as subjective, or observer-dependent quantities (e.g., Shannon's information "entropy") where the "entropy" of a system depends on what an individual knows about it. The reader should be aware that some authors illegitimately conflate these meanings. In the present paper, to be clear, the word entropy is used in its physical thermodynamic sense as defined.

REFERENCES

Alexander, R. 1971. *Size and Shape*. London: William Clowes & Son.

Aristotle. 1947. "De anima." Pp. 163-329 in *Introduction to Aristotle*, edited by R. McKeon. New York: Random House.

Bertalanffy, L. von 1952. *Problems of Life*. London: Watts.

Boltzmann, L. 1974 [1886]. "The Second Law of Thermodynamics." Pp. 13-32 in *Theoretical Physics and Philosophical Problems*, trans. by S.G. Brush. Boston: D. Reidel Publishing.

Bramwell, A. 1989. *Ecology in the 20th Century: History*. New Haven, CT: Yale University Press.

Callebaut, W., and R. Pinxten, eds. 1987. *Evolutionary Epistemology: A Multiparadigm Program*. Dordrecht: D. Reidel Publishing.

Calvin, W. 1986. *The River That Flows Uphill: A Journey from the Big Bang to the Big Brain*. New York: Macmillan.

Campbell, D.T. 1987. "Evolutionary Epistemology." Pp. 47-89 in *Evolutionary Epistemology, Rationality, and the Sociology of Knowledge*, edited by G. Radnitzky and W.W. Bartley, III. La Salle, IL: Open Court.

Carneiro, R. 1970. "A Theory of the Origin of the State." *Science* 169: 733-738.

_____. 1987. "Village Splitting as a Function of Population Size." Pp. 94-124 in *Themes in Ethnology and Culture History: Essays in Honor of David F. Aberle*, edited by L. Donald. Meerut: Archana Publications.

Cloud, P. 1988. *Oasis in Space: Earth History from the Beginning*. New York: Norton.

Darwin, C. 1937 [1859]. *On the Origin of Species by Means of Natural Selection or the Preservation of Favored Races in the Struggle for Life*. New York: D. Appleton-Century.

Dawkins, R. 1982. *The Extended Phenotype*. San Francisco: Freeman.

_____. 1986. *The Blind Watchmaker*. New York: Freeman.

Dennett, D. 1995. *Darwin's Dangerous Idea*. New York: Simon & Schuster.

Depew, D., and B. Weber. 1995. *Darwinism Evolving*. Cambridge, MA: MIT Press.

Descartes, R. 1975 [1644]. "Conservation of Quantity of Motion." Pp. 234-266 in *Energy: Historical Development of the Concept*, edited by R. Lindsay. Benchmark Papers On Energy, Vol. 1. Stroudsburg, PA: Dowden, Hutchinson & Ross.

_____. 1986 [1637]. *Discourse on Method and Meditations*, trans. L. Lafleur, trans. New York: Macmillan.

Dyke, C. 1997. "The Heuristics of Ecological Interaction." Pp. 49-97 in *Advances in Human Ecology*, Vol. 6. Gereenwich, CT: JAI Press.

Eddington, A. 1928. *The Nature Of The Physical World*. Ann Arbor, MI: Ann Arbor Paperbacks.

Fisher, R.A. 1958 [1930]. *The Genetical Theory of Natural Selection*. New York: Dover.

Gibson, J.J. 1986. *The Ecological Approach to Visual Perception*. Hillsdale, NJ: Lawrence Erlbaum.

Gould, S.J. 1989. *Wonderful Life: The Burgess Shale and the Nature of History*. New York: W.W. Norton.

Kim, N., M.T. Turvey, and C. Carello. 1993. "Optical Information About the Severity of Upcoming Collisions." *Journal of Experimental Psychology* 19(1): 179-193.

Koestler, A., and J.A. Smythes, eds. 1969. *Beyond Reductionism: New Perspectives in the Life Sciences*. New York: Macmillan.

Lange, F. 1950 [1877]. *The History of Materialism*. New York: The Humanities Press.

Lee, D. 1980. "A Theory of Visual Control of Braking Based on Information about Time-to-Collision. *Perception* 5: 437-459.

Leibniz, G.W. 1925 [1696]. "Explanation of the New System of Communication Between Substances, by Way of Reply to What is said about it in the Journal of September 12, 1695." Pp. 319-330 in *The Monodology and other Philosophical Writings*, trans. R. Latta. London: Oxford University Press.

_____. 1953 [1714]. "Monodology." Pp. 249-272 in *Leibniz*, trans. G. Montgomery. La Salle, IL: Open Court Publishing.

Levins, R., and R. Lewontin, R. 1985. *The Dialectical Biologist*. Cambridge, MA: Harvard University Press.

Lewontin, R. 1992. *Biology as Ideology: The Doctrine of DNA*. New York: Harper Collins.

Lorenz, K. 1973. *Behind the Mirror: A Search for a Natural History of Human Knowledge*. New York: Harcourt Brace Jovanovich.

Margulis, L., and J. Lovelock. 1974. "The Biota as Ancient and Modern Modulator of the Earth's Atmosphere." *Tellus* 26: 1-10.

Maruyama, M. 1963. "The Second Cybernetics: Deviation-Amplifying Mutual Causal Processes." *American Scientist* 51: 164-179.

Maynard Smith, J. 1988. "Evolutionary Progress and Levels of Selection." Pp. 219-230 in *Evolutionary Progress*, edited by M. Nitecki. Chicago: University of Chicago Press.

Mayr, E. 1980. "Prologue: Some Thoughts on the History of the Evolutionary Synthesis." Pp. 1-48 in *The Evolutionary Synthesis*, edited by E. Mayr and W.B. Provine. Cambridge, MA: Harvard University Press.

_____. 1985. "How Biology Differs from the Physical Sciences." Pp. 43-63 in *Evolution at a Crossroads*, edited by D. Depew and B. Weber. Cambridge, MA: MIT Press.

Munz, P. 1985. *Our Knowledge of the Growth of Knowledge: Popper or Wittgenstein?* London: Routledge & Kegan Paul.

_____. 1987. "Philosophy and the Mirror of Rorty." Pp. 345-387 in *Evolutionary Epistemology, Rationality, and the Sociology of Knowledge*, edited by G. Radnitzky and W.W. Bartley, III. La Salle, IL: Open Court.

Peck, A.J. In Press. "Hydrodynamics, Reactive Forces, and the Autocatakinetics of Hearing." *Ecological Psychology.*

Popper, K. 1985. *Unended Quest: An Intellectual Autobiography.* La Salle, IL: Open Court.

Prigogine, I. 1978. "Time, Structure, and Fluctuations." *Science* 201: 777-785.

Radnitzky, G., and W.W. Bartley, III, eds. 1987. *Evolutionary Epistemology, Rationality, and the Sociology of Knowledge.* La Salle, IL: Open Court.

Rescher, N. 1967. *The Philosophy of Leibniz.* Englewood Cliffs, NJ: Prentice-Hall.

Schröedinger, E. 1945. *What is Life?* New York: Macmillan.

Schwartzman, D., S. Shore, T. Volk, and M. McMenamin. 1994. "Self-Organization of the Earth's Biosphere-Geochemical or Geophysiological?" *Origins of Life and Evolution of the Biosphere* 24: 435-450.

Schweber, S. 1985. "The Wider British Context of Darwin's Theorizing." Pp. 35-69 in *The Darwinian Heritage*, edited by D. Kohn. Princeton, NJ: Princeton University Press.

Sober, E. 1984. *The Nature of Selection: Evolutionary Theory in Philosophical Focus.* Cambridge, MA: Bradford Books.

Spencer, H. 1852 [1857]. "The Social Organism." *Westminster Review* 57: 468-501.

_____. 1862. *First Principles.* London: Williams and Norgate.

_____. 1892 [1857]. "Progress, Its Law and Cause." Pp. 8-38 in *Essays: Scientific, Political, and Speculative.* New York: D. Appleton and Company.

Swenson, R. 1988. "Emergence and the Principle of Maximum Entropy Production: Multi-Level System Theory, Evolution, and Non-Equilibrium Thermodynamics." Page 32 in *Proceedings of the 32nd Annual Meeting of the International Society for General Systems Research..*

_____. 1989a. "Engineering Initial Conditions in a Self-Producing Environment." Pp. 68-73 in *A Delicate Balance: Technics, Culture and Consequences*, edited by M. Rogers and N. Warren. (IEEE Catalog No. 89CH2931-4). Los Angeles: Institute of Electrical and Electronic Engineers.

_____. 1989b. "Emergent Evolution and the Global Attractor: The Evolutionary Epistemology of Entropy Production Maximization." *Proceedings of the 33rd Annual Meeting of the International Society for the Systems Sciences* 33(3): 46-53.

_____. 1989c. "Emergent Attractors and the Law of Maximum Entropy Production." *Systems Research* 6:187-1987.

_____. 1989d. "Gauss-in-a-Box: Nailing Down the First Principles of Action." *Perceiving-Acting Workshop Review* (Technical Report of the Center for the Ecological Study of Perception and Action) 5: 60-63.

_____. 1990. "Evolutionary Systems and Society." *World Futures* 30: 11-16.

_____. 1991a. "End-Directed Physics and Evolutionary Ordering: Obviating the Problem of the Population of One." Pp. 41-60 in *The Cybernetics of Complex Systems: Self-Organization, Evolution, and Social Change*, edited by F. Geyer. Salinas, CA: Intersystems Publications.

_____. 1991b. "Order, Evolution, and Natural Law: Fundamental Relations in Complex System Theory." Pp. 125-148 in *Cybernetics and Applied Systems*, edited by C. Negoita. New York: Marcel Dekker.

_____. 1992. "Autocatakinetics, Yes—Autopoiesis, No: Steps Towards a Unified Theory of Evolutionary Ordering." *International Journal of General Systems* 21(2): 207-228.

_____. 1996. "Thermodynamics and Evolution." In *The Encyclopedia of Comparative Psychology*, edited by G. Greenberg and M. Haraway. New York: Garland.

_____. In Press-a. "Spontaneous Order, Evolution, and Autocatakinetics: The Nomological Basis for the Emergence of Meaning." In *Evolutionary Systems*, edited by G. van de Vijver, S. Salthe, and M. Delpos.

_____. In Press-b. "Evolutionary Theory Developing: The Problem(s) with 'Darwin's Dangerous Idea.'" *Ecological Psychology*.

_____. In Press-c. *Spontaneous Order, Evolution, and Natural Law: An Introduction to the Physical Basis for an Ecological Psychology*. Hillsdale, NJ: Lawrence Erlbaum.

Swenson, R., and M.T. Turvey, M.T. 1991. "Thermodynamic Reasons for Perception-Action Cycles." *Ecological Psychology* 3(4): 317-348.

Turner, J., and A. Maryanski. 1979. *Functionalism*. Menlo Park, CA: Benjamin/Cummings.

Turvey, M.T., and R.E. Shaw. 1995. "Towards an Ecological Physics and a Physical Psychology." Pp. 144-169 in *The Science of the Mind: 2001 and Beyond*, edited by R. Solso and D. Massero. Oxford: Oxford University Press.

Vernadsky, V. 1986 [1929]. *The Biosphere*. London: Synergetic Press.

Williams, G. 1992. *Natural Selection*. Oxford: Oxford University Press.

THE HEURISTICS OF ECOLOGICAL INTERACTION

C. Dyke

ABSTRACT

Human ecology obviously needs to be linked closely to economics. But present-day economics remains dominated by the theory and heuristics characteristic of the physics of two or three centuries past. This was a physics of two-body problems, where linear approximations were nicely adequate. But systems of complex interactions are not adequately explained or understood as two-body problems. Linear approximations can't be expected to succeed. "Standard" statistical strategies have to be supplanted by new more sophisticated ones.

The natural sciences have recognized the situation and have moved on to a genuine confrontation of nonlinearity, interactivity, and both stability and instability far from equilibrium. This paper urges economists to do likewise, and tries to show why they must.

Advances in Human Ecology, Volume 6, pages 49-74.
Copyright © 1997 by JAI Press Inc.
All rights of reproduction in any form reserved.
ISBN: 0-7623-0257-7

INTRODUCTION

The fundamental picture of the world inherited by Western science as it entered its period of incredible growth and development in the modern era was a dualistic one. On the one hand, there are human subjects, sovereigns of their fate; and on the other hand, there are "physical" or "natural" objects, raw material for human fulfillment. This is, of course, a Biblical inheritance, as old as the story of the creation, where Adam is given dominion over all. I think that the dualism of "man the master" and "nature the material" is the historically fundamental dualism, and not the dualisms of soul and body, spirit and matter, or mind and body that seem fundamental only as they function as ideological justification for our possession and control of the planet earth.

As we grapple, here at the end of the twentieth century, with our relationship to the environment, it's no surprise that our attempts to "solve environmental problems" are still dominated by this old dualism. But it's not working out very well. It appears that far from being sovereigns over "nature," we're deeply embedded in it. Our relation to "nature" is interactively complex. The styles of doing science that grew up in the heady days of our presumed sovereignty are inadequate to deal with interactive complexity. However, dialectically, the sciences themselves have evolved tools that may be adequate. My view is that these tools are now available, but we're not yet using them with skill. We need to get rid of the old heuristic habits of the old style of doing science.

To remind ourselves of these old habits, we don't need to rehash the whole history of Western science. We need only begin with the recent past and recall that the positivist caricature of scientific "activity" was that laws or law-like propositions were stated as regularities, and then these regularities were tested for reliability—at least at some statistical threshold. If the research reports in *Science* and *Nature* over the last couple of years are a reliable cross-section of what goes on in science these days, activity on the positivist formula is very rare. This is not to say, of course, that hypotheses are not tested. They are all the time. But hypotheses are virtually never laws, and very rarely derived from laws on the model the positivists liked to believe in. And this is not to say that laws aren't still being discovered now and again, or that tests of laws have disappeared from science. It's just that this part of scientific activity is far from representative—from a statistical point of view, let us say.

The range of experimental heuristics in current science is enormous. An ever more elaborate division of scientific labor produces ever more sophisticated and intricate strategies and investigative tools. Among the tools one finds used these days, when they are appropriate, are the tools of nonlinear dynamics and nonequilibrium thermodynamics, new but improving rapidly. These tools are used because more and more of the problems that attract scientists involve nonlinear and/or far from equilibrium phenomena. The heuristics required for the use of these are enormously complex and ingenious.

My intention is to provide enough of the heuristics we need so that by the time we get to the end of this paper, we can begin to see fairly clearly how to go about understanding the thoroughly interactive system we call human ecology. We will find that the investigative schemes we need emphasize characteristic patterns rather than laws (though laws of a very fundamental sort lie behind the patterns). We will also find plenty of contingency along with limits on predictability at the level of our interchange with our environment. Old-style science would have considered this a roadblock to understanding. We'll see that we don't need to think that way at all. We'll be able to trade Leibnizean certainties for practical understanding.

"LINEAR" NONADDITIVITY

We can start by trying to get a basic understanding of nonlinearity. Actually, this is an issue totally familiar to economists, who have long since learned to deal with the nonadditivity of utilities. Thus, we can start on familiar ground and get our bearings for an expansion of our viewpoint. The utility theory that finds its current place in game theory, price theory, and, say, the theory of consumer surplus had its beginnings in Hume, Smith, and Bentham. At the beginning, it was confidently expected that utilities could be added up and compared, and now everyone has accepted the fact that this can't be done. Instead, the calculations involved in dealing with utilities must be consistent with the fact that utilities are unique only up to a linear transformation. In consequence, utility rankings are defined as interval rankings. What this means, from one point of view, is that utilities have no natural quantitative scale.[1] There is no metric unit whose adoption allows cardinal expressions of quantity. In the absence of such a measure, there's no way that utilities can be added up, in general.[2] There's nothing intrinsically devastating in this state of affairs. Temperatures, for example, can't in general be added up even when expressed on the Kelvin scale. In the absence of a natural scale of utilities, economists have come to work with preference rankings and the ratios established in auctions and pseudo-auctions. In many local circumstances, this is sufficient to produce useful results, results we can summarize as a coherent theory of price. Typically, attempts to extend local results to more global contexts produce severe problems, especially problems of commensurability. Some of the most serious issues of our times involve these problems—for example, environmental issues, where the attempt to "internalize" environmental values (that may intrinsically be gross externalities) has so far failed.

Bracketing such global issues, let's look more closely at how we work with utilities in the absence of a cardinal metric. The first thing we notice is that we "cheat" a lot. We use things like dollars and florens as cardinal measures. We treat prices as if they were cardinal, though we know they're not. How do we get away with that? Well, there's really not much to get away with under the right circumstances. The trick is to establish contexts within which quantities of various things (including money, most often) are sufficiently connected to establish a common scale. And the usual trick is to assume a system of equilibrium prices.[3] Examining this assumption, a bit later, will allow us to link up the nonadditivity of utilities with other nonadditivities. The concept of a system at or near equilibrium will, of course, be central. At the moment, however, we have to go back to the interval rankings that are the actual stuff of utility theory. The claim that utilities are unique only up to the linear transform $U_2 = aU_1 + b$, where a and b are constants reminds us that the standard techniques of game theory (and the equivalent segment of Bayesian decision theory) are the manipulations of the standard matrices of linear algebra. For example, the linear transform, suitably well defined, is used all the time to get zeroes where we need them in the matrices for various reasons.

Such manipulations are equivalent to coordinate transformations that preserve local topography. Any system defined by a matrix "lives" on a surface with a local topography. The effect of linear transformations is to leave the local properties of the surfaces unchanged. If the local topographical direction from a given point to another is "up," then it remains "up" on the transformed surface; if "down," it remains "down;" and if "flat," remains "flat." The surface can stretch and contract uniformly while these properties remain unchanged, and any set of coordinates can be moved around to let (0,0) reside anywhere we please. When we think of matters this way, it's easy to see why we can say that such systems have no natural scale.[4] We can also reflect that one of the main differences between classical and quantum physics is that quantum theory asserts that there *is* a natural scale to phenomena, indicated by Planck's constant and its variants, including the Planck distance.

Another way to look at this situation is to see that with linear transformations at no preferred scale, everything moves independently of everything else, but everything moves in step. Everything grows or shrinks independently of everything else, but everything grows or shrinks the same amount. The surface so characterized is extremely well behaved, and it is this good behavior that allows us, for example, to treat individual demand schedules as (a) commensurate and (b) aggregatable into a coherent system of prices. That is, we confidently expect one more passenger on the bandwagon to add an incremental load proportional to the sum of potential passengers—in the limit and, at equilibrium, an infinitesimally small increment.

One of the best ways to see how our normal intuitions and habits are formed by assumptions of linearity of this sort is to see that such assumptions are canonized in standard liberal democratic theory. First, everyone is to count as one, and not more than one, as Mill put it in defence of the democratic core of his theory of utility—or, one person, one vote, as we put it in terms of voting rights. Each of these voters is assumed to be autonomous. That is, any change in the vote of one person is independent of that of every other person. This insures the required linearity. As individual votes change, the social decision will move across the landscape point by point continuously and smoothly. Weight of numbers, where each number has a marginal weight, will rule.

It is easier than we might like to think to violate the conditions for such additivity. All we need to do is introduce interdependencies among alternatives and/or voters. For example, we can think of a system of voters faced with a binary decision between two alternatives that are internally complex. That is, each alternative is made up of a "market basket" of positions on a number of issues. We can also imagine a process of coalition formation in which voters tie themselves to one another with respect to the two alternatives. Obviously, coalitions could form with respect to any of the individual issues within the market basket. In short, we imagine a situation which may be absolutely the norm in, say, modern two-party democracies but which may not yield a well-behaved voting surface. I have argued elsewhere, in effect, that reducing a decision problem to a binary choice is best understood as a default move for cases in which the conditions for equilibrium are unattainable (Dyke 1983).

Indeed, a good deal of liberal theorizing about democracy has been directed toward the perceived problems for democratic autonomy resulting from interdependencies, and the natural tendency for coalitions to form in order to deal with them. The classic American work in this vein is Madison's contribution to *The Federalist*. The attempt is often made to rule out coalitions and other interdependencies, though what it would mean to "rule them out" is extremely problematic. It doesn't do very much good to be told that such and such an idealized "citizen," cousin to the "rational economic man," would avoid the problematic interdependencies.

As committed democrats, should we expect that the conditions for individualism and additive collectivity will be inexpensive and easily forthcoming, or should we expect that constant work will have to be done in order to preserve those conditions? Madison's answer here, and in fact the answer of almost every liberal theorist, is that work will have to be done to maintain the conditions of independence. They wish it were otherwise. They have sad stories to tell about the lamentable failures. But they are certain that the natural tendency is for structure to emerge. The new understanding of nonlinear systems gained over the last few years underwrites the wisdom of this judgment while casting doubt on the contrary judgment made far too often in orthodox economics.

So, there is something definitely paradoxical about saying that the shift away from linearity as the norm is something new, a paradigm shift. The classic works in liberal political theory argue that additive individualism ought to be the norm, but they know that their ideals are under constant threat from the normal course of events. The example serves as an interesting reference point when we need a concrete image of potential violations of linearity. With it in mind, we can dig a little deeper.

A FIRST EXTENSION

The range and utility of linear algebra goes far beyond the treatment of strictly linear functions or relations. The key to the extent of the range of linear algebra is the familiar postulate that any short smooth curve can be approximated by a straight line. The ability to avail oneself of the approximation then boils down to adjusting *how* short, *how* smooth, and *how* close the approximation needs to be. Enormous numbers of applications in which appropriate degrees of length, smoothness and closeness can be assured make linear algebra one of the most ubiquitous mathematical tools of many disciplines. In general, the ability to avail oneself of linearized approximations depends on the confidence with which one can specify the effects of increments in variables, especially the effects of variables upon themselves. The long traditional infatuation with ballistics trajectories and their close analogues fostered inaccurate habitual expectations among us, but to say this is certainly not to say that linearization strategies are completely obsolete, or inapplicable in the context of nonlinearity.

It's relatively easy to move from reflections on linearity and matrices to linearities and vector spaces. Linear spaces are defined in terms of the additive and multiplicative properties of vectors. So the question of the additivity of utilities is, indeed, a typical one. In any single dimension of a linear space, vectors can be resolved in such a way that a single resultant vector can be found that is the consequence of some number of additions and multiplications performed on all the vectors in that dimension. Vectors in any one dimension are said to be dependent, and vectors in different dimensions are said to be independent. To put this in slightly different terms, the vector behavior of a single degree of freedom can always be expressed as a single vector. Linear spaces are thus *flat* in a straightforward sense. Coordinate tranformations are, again, smooth, and if we choose, we can represent the vectorial portrait at any point in the space by means of *n* vectors lying on the *n* axes. Furthermore, the flatness of the transformations allows us to place the tails of all the vectors at the origin of the axes. The equivalence of this to the matrix manipulations discussed above should be obvious, for once again we are dealing with coordinate transformations.

The additive and multiplicative properties of vectors in linear spaces are the defining properties of such spaces. This is why people urging us to consider nonlinear systems are so inclined to identify nonlinearity and nonadditivity. And this turns out to be a very fundamental feature of a very particular kind of mathematical space. We then have to reflect on the properties of these spaces and their appropriate use.

What kind of dynamical systems can live in a linear space? Basically, linear space is the home of systems of independent "forces" that can be separated out in such a way that their individual contributions to the trajectory of the system can be examined by themselves, with small increments of "force" producing small changes in trajectory, and we can say clearly what the trajectory would have been had a given individual "force" not been operating. A good example for economists of an explicit attempt to confine a system to linear space is Arrow's axiom of the Independence of Irrelevant Alternatives in the famous 1951 proof. The effect of the axiom is to insure that if an element in a weak ordering is deleted, all the other elements, placed in a new ordering, will retain their relative positions with respect to one another that they had in the old ordering. Alternative objects of choice are, in a word, supposed to be independent of one another. Or, to put things in a more dynamical way, any "force" moving a chooser toward an object of choice m is independent of any force moving the chooser toward an object of choice n. The relative preference for m is independent of the presence or absence of n in the field of alternatives. This example, of course, links, yet again, game theory, normal utility, and linearity.

In summary, thoroughgoing linearity seems the most natural thing in the world as it appears in standard game theory. It conforms absolutely with the ideology of individual independence. Similarly, it's essential for the overriding goal of finding equilibrium solutions, either point attractors or probabilistic strategies. In both these respects, game theory stands in the same position as all the other classic liberal methodologies—virtually the whole of Western social science including economics. Hence, it can stand *for* them as an exemplar. For example, when we look at "high-powered" econometric forecasting models with their hundred-plus equations and hundred-plus parameters, we see complication and vastly increased computational difficulty, but the basic terms of the model-building logic have not changed. Convergence is still obtained by wholesale linearization.

It may seem too basic to need saying, but calculus is through and through a method of linearization and, for purposes of integration, space smoothing. And we should not need to be reminded that the most common technique for linearization is to find an expansion for an expression and throw away higher-order terms. My own interest in nonlinear dynamics began when I heard a lecture by Dick Levins in which he pointed out that the higher-order terms were interaction terms and that throwing those terms away was tantamount to deciding that the interactions within a system do not contribute significantly to its dynamics. But in many systems, interactions dominate the dynamics, and in any case, no system sensitive to initial conditions can be expected to conform to the pseudo-approximation of its linearization. This point is now firmly established in the hard sciences and must become established in those areas such as human ecology that are, in a plain sense, *all about* interactions (Young 1996).

Systems that live in linear spaces are characteristically systems where no interactions take place. A system in which "forces" influence and alter the conditions of their own exercise lives in nonlinear spaces. These are systems where an examination of coefficients is not sufficient to predict the behavior of the system. In particular, what we might call *importantly* nonlinear systems are ones in which the normal expansions do not approach a limit fast enough for higher-order terms to be neglected in computation. These are the now-well-known systems infinitely sensitive to initial conditions, the systems for which no linear approximation is good enough to capture the dynamics. One of the ways in which this last situation is described is by saying that the surfaces on which nonlinear dynamical systems live are stretched and folded by the interactive dynamics of the system itself.

Another good way to contrast linear and nonlinear systems is by thinking of the way in which they "explore the spaces they live in." Linear systems characteristically follow a relatively simple single trajectory, or settle down, after a suitable period of time, at a single point. Nonlinear systems characteristically do neither. The chaotic or strange attractors that characterize their assymptotic behavior are, in effect, patterned explorations of the *whole* of the space available to them. (This is what is meant by ergodic behavior.) Among the more important consequences of this patterned exploration is that *averaging* over any period of time is a tricky business and, in general, obscures understanding of the system rather than contributes to understanding. From the point of view of the normal habits of the economist, this means that standard regression techniques are almost certain to be completely useless. The search for trends thus will not be a very rewarding activity except in the very short run.

Ergodicity is itself, of course, a statistical concept. That is, if a chaotic attractor is ergodic, then we can expect that an iterated point will visit any given region of the attractor as often as it visits any other. But this doesn't mean that normal statistics can handle ergodicity, for the attractor itself is a very complex mathematical object, and a point visiting its regions uniformly will, in general, visit the regions of space the attractor is embedded in very un-uniformly.[5] The situation is complicated by the fact that chaotic orbits are interlaced with periodic orbits infinitely densely.

The upshot of this complexity is felt acutely in any study of weather and climate. The rhythms of the seasons, rainy and dry, hot and cold, are *sort of* periodic, as are many other natural rhythms. "Sort of periodic" is almost always treated as "almost periodic," but in the dynamical bundle of climates "near" the one we have, for every periodic one there are many more that are not periodic at all. In the final section of this paper, we'll have a closer look at what this means in terms of policy.

NON-EQUILIBRIUM PROCESS

We now move to the second of the two concepts later to be combined. We'll look at some rudimentary features of linear non-equilibrium thermodynamics. We begin with linear non-equilibrium models because they stand in the middle between equilibrium models and nonlinear models of systems far from equilibrium. They offer the chance to see fairly clearly what's at stake in the distinction between linearity and nonlinearity. In particular, we'll be able to see behind the mathematics itself to the differences in the *kinds of processes* characterized as linear or nonlinear, and the kinds of expectations we ought to have about these processes.

Thermodynamics is a species of dynamics whose primary intended use is to discover the likely—or, in the best case, necessary—future trajectory of systems characterized in terms of state variables such as temperature, volume, and pressure. Most of thermodynamics is primarily suited for examining the behavior of systems at or very near equilibrium. In fact, many key thermodynamic terms (such as "entropy") have their only strict definitions at or near equilibrium. By the same token, many key thermodynamic quantities (such as temperature) can be *measured* only at or near equilibrium. So, if strictly quantitative results are sought for a system, that system's relation to equilibrium is an absolutely primary concern.

Unfortunately, the only systems that can be reliably expected to be at equilibrium are isolated systems (after a suitable transient period) and things that are at the bottom of very deep energy wells. By definition, equilibrium is boring. Fortunately, it can also be beautiful, as in a diamond. But in general, interesting things are neither isolated systems nor at the bottom of deep energy wells, so the search for ways to extend the resources of thermodynamics to interesting things has been a continuous one over the last century. There have been interesting qualitative results and, a bit less frequently, interesting quantitative results.

Think of some system that is maintaining itself in a steady state while irreversible processes are going on in and around it. In particular, think of a sort of continuous flow process in which a synthesis is taking place in what we'll call "the chamber" in the presence of a catalyst. The substrates flow in, the catalysis occurs, and the product flows out. We set the process up so that there is a vanishingly small probability that the reverse reaction will take place. In order to understand the system, perhaps to come to viable hypotheses about the nature of the catalytic system, we would want some quantitative results. For practical experimental reasons, this very often means that figuring out what's going on in the chamber requires taking measurements outside the chamber.

Taking measurements outside the chamber and relating them to what is going on inside requires a long chain of inferences and assumptions, each of which has to be tested experimentally. What the researcher would like to find is a smooth curve relating state variables at equilibrium with state variables away from equilibrium. This relation has to be *fought for*, inch by inch, case by case. It can not be blithely and confidently assumed, so it's impossible to produce the narrative of such a battle here.[6] But that's exactly the moral of the story. Experimental design and measurement strategies far too complicated (and sometimes arcane) to appear in a paper like this one are required in order, say, to make experimental inferences about the relationship of a relatively simple membrane system *at* equilibrium and the same system in its operating range *away* from equilibrium. Understanding the dynamics of biological membranes thus has taken some very cleverly prepared experimental systems and sophisticated measurement techniques. In contrast, the standard practice of economists is to "accomplish" the equivalent task with a shrug and a wave of the hand. When natural scientists sneer at economics as bad science, it is this sort of practice (among others) they have in mind.

We can thus expect that the move into the realm of nonlinear systems stabilized far from equilibrium is far from easy. Fortunately, a lot of clever work has been done here also. We can sample this work by considering the project of Rod Swenson and his colleagues (e.g., Swenson 1989). Like many others these days, Swenson is looking closely at systems with nonlinear dynamics operating far from thermodynamic equilibrium. His aim is to produce a general theory of evolution grounded in physics and to account for the emergence of complexity as a natural expectable phenomenon rather than as something surprising. At this point, it seems to me that the foundational insights anchoring his project are better than the corresponding insights of some of the competing projects, although one would hope that convergence and consensus would begin to appear fairly soon.

The first of Swenson's foundations is the "law of maximum entropy production." Just as we know that a conservative system is somewhere on the constant energy surface, so we know that a dissipative system is on a trajectory of maximum entropy production. That is, a system will be in a state that maximizes the rate at which the rate of entropy increase increases. So, where S is entropy, and $P = dS/dt > 0$, then $dP/dt > 0$ is at a maximum.

From one point of view, Swenson's law of maximum entropy production is the successor to a number of earlier attempts to find some sort of entropy invariant. It turns out that the law of maximum entropy production is far more straightforward than the history of attempts to formulate it would suggest, and even more straightforward and intuitively comprehensible than, say, the law of the conservation of energy or the principle of natural selection. In its raw form, it simply lets us understand why the toilet will flush away our waste *as fast as it can* and gives a precise definition to "as fast as it can." Or, equivalently, if we ask why the water forms a vortex as it goes down, Swenson's law gives us the framework we need for an explanation. It says: "To understand the vortex, examine the system for gradients, dissipative surfaces, and constraints. The explanation of the vortex is the account of the maximal use of the dissipative surfaces maximally accessible to the system under the given constraints to eliminate the gradients as quickly as possible."

Notice that Swenson's law is not by itself an explanation of anything. Like the principle of the conservation of energy for conservative systems, and the principle of natural selection within its bounded explanatory sphere, Swenson's law is a "recipe" for explanations and a guide to formulating them. Like the other two "laws," it can even have the distinct smell of tautology about it when it is expressed generally. For, like all laws, it claims that it couldn't fail to hold for the phenomena for which it holds. Yet, it has the same power as other laws of its generality in that it determines viable candidates for successful explanation, and these explanations are *not* tautological. Notice also that in the case of the toilet, as in all other cases for which Swenson's law is centrally relevant, the phenomenon to be explained is not start-specific. The vortex is not to be explained on the basis of the initial state of the water and waste but on the basis of its path to an end point—the elimination of all gradients in as brief a time as is consistent with the constraints. I've discussed "finish-specificity," characteristic of systems thought of in thermodynamic terms, elsewhere (Dyke 1994b). The idea is to contrast them with "start-specific" systems, characteristic of easily linearizable trajectories confined to a single path. Finish-specific systems collect many different starting points whose trajectory ends at the same specifiable attractor state.

An important next step is to see that the journey a system takes to the elimination of gradients is not a passive one. A system actively explores opportunities for dissipation. "Actively explores?" The exorcist here sniffs the sulphur of anthropomorphism, but my advice, if you remember, is to ignore the exorcists. There are enormous advantages in having a language of "active seeking" available for finish-specific systems, and these advantages far outweigh the disadvantages, especially as we leave behind our old bad habits. After all, we have been (usefully) telling each other for a long time that water seeks its own level without attributing any deep psychological capacities to water. We are simply emphasizing the finish-specificity of a process as a key to understanding it.

A very simple example will help get some of these ideas straight—just about the simplest dynamical system one can imagine.[7] In this system, a new x is gotten from an old one simply by doubling the old one: $x_n = 2x_o$, where we "feed" each new x back into the equation. That is, we iterate the function. However, the system is confined to the unit interval, so every x has to lie between 0 and 1. There are two ways to think about this. Either you say that $x_n = 2x_o$ modulo 1, so that when $2x$ is more than 1 you throw away all but the fractional part of the product; or, more illuminating, think of the whole process going on around a circle one unit in circumference. Then, multiplying x by 2 takes you some part of the way around the circle or, sometimes, more than part way so that you pass the starting point. So, for example, the point 2/3 maps onto the point 4/3, but 4/3 is located at the point 1/3. Similarly, the point 1/6 maps onto the point 1/3. Iterating this latter again, 1/3 goes to 2/3, and again, 2/3 goes to 4/3, and here we are back at 1/3. The point 1/3 gathers the starting points 1/6 and 2/3 and puts them into a tight loop: 1/3, 2/3, 1/3 and so forth, over and over again. But, as you can see, all starting points $1/3n$ where n is even will be gathered in the same loop.

The first point is to see that this process turns out to be finish-specific in a particular sense. Given the mapping function, there will be a family of inevitable finishing patterns, each having its own little family of starting points it collects. But further, if we notice the system behaving in a certain way, that is, 1/3, 2/3, 1/3, 2/3 over and over, there is no way we can determine the initial starting point that led to the loop.

Well, maybe that's just annoying, but suppose we made a machine out of this mapping. To run the machine forward, we just put in a starting x, and let the machine iterate until it gets to one of the finish-specific end states. But can we get the machine to run backwards to recapture the starting point? Clearly not. Start it backwards and, I suppose, it just begins to divide by 2. It keeps getting an answer that is $1/3n$. But where is it supposed to stop? It has no idea. And further, in going back just by dividing 1/3 by 2, it may never reproduce the reverse of the process that got it to its end state in the first place. Suppose, for example, that the initial x was 2/3.

Most end-specific systems are irreversible in the way this simple system is. They can't be shoved into reverse and be expected to retraverse the pathway that got them where they are. And that, of course, isn't just a mathematical curiosity, for it drives right to the heart of our naive feeling that natural systems ought to be able to fix themselves when things go wrong. Only if they are operating in a linear range very near equilibrium can we hope that they'll right themselves, for in that range they can find their way back—and then only if what went wrong is extremely minor. We've just seen an example of a system that can't find its way back even though it is so constricted that it can explore only a tiny (though infinite) part of its phase space.

Once we're willing to talk about a system exploring its available opportunities for dissipation, we are finally able to assess the possible futures of the system with the right intuitive picture in mind. In particular, we ready ourselves for the next important step, which is to see that organized opportunities for dissipation *emerge* in the course of the history of the system. These emergent opportunities are always associated with—indeed, are tantamount to—the emergence of relatively stable spatio-temporal structures. Swenson himself has worked this through with Bénard cells, a supersolid example where the emergence of the cells is very easy to understand on the basis of accessing dissipative opportunities, and impossible to understand otherwise. Perhaps an even more fundamental "example" is to reflect that conduction, convection, and diffusion are three basic access routes to the elimination of gradients, they operate at different scales, and the availability of each route depends on organizational features of the system within which it occurs, however minimally organized these features might be—in the case of conduction, *very* minimal.

Swenson is involved in an overall project to extend this basic picture over the spectrum of all evolving systems, including social systems. Much work obviously remains to be done, but it is worth mentioning a general consideration that becomes crucial as we begin to assess the possible futures of social systems and how we model them. The ways in which any two "things" can interact depend upon how they can, in a broad sense, "communicate" with one another. This is vividly illustrated in the fundamental extreme when we learn how to understand light cones in space-time as a representation of the boundary between events that can affect one another and those that can not. But there are other limits. Neutrinos, for example, are extremely uncommunicative and extremely rarely combine with anything else.

Anything decently complex has innumerable features that are potential sites of communication. For example, *in vivo* DNA is in constant interaction with protein complexes to open and close various potential sites of interaction. Proteins themselves have activated and inactivated configurations. This can be looked at (and has been looked at) in terms of variable and differential information exchange. Now, the "communication" between macromolecules is intensely connected to the energetics of interaction between them. Not all communication is so tightly coupled to energetics. Take, for example, the information we gain from something by looking at it. Given the solar source of the reflected photons we utilize, we get essentially a free peek (or "peak," spikily speaking). We just interrupt the trajectory of a few photons and give them a sink a bit different than the one they would otherwise find. And there are, in general, plenty of photons to go around. Things don't wear out by being looked at. Yet, visual access to something can potentiate it for all sorts of other interactions that are far more coupled to its energetic economy, making it available for integration into all sorts of structure including social structure. Swenson and his colleagues have woven such considerations into a basically Gibsonian framework with all sorts of fascinating prospects on the horizon.

Given the connection between communication and dynamics, it isn't surprising that every economic theory, including the most orthodox, defines the system of communicative interaction that it is willing to consider—hence, defines the range of potential consequences of communicative interactions. In fact, orthodox economics is extremely strict in its definition of communicative access with, for example, "perfect information" assumptions at key junctures, restrictions on price information at others, and so forth. The specification, and especially restriction, of communicative interaction within orthodox economic theory is yet another of the devices designed to ensure that the trajectory of an economy is confined to a linear range. Precisely the information that would lead to the characteristic behavior of nonlinear systems—the emergence of structure—is barred. But while the economists can prevent their theories from considering such information, they can't prevent real economic agents from doing so. Real agents communicate "on many channels," thus enriching the system of interactions. In consequence, a space of possible self-organizations is explored that orthodox economic theory scruples to ignore.

From the point of view of human ecology, the most important consequence of the organization emergent in the economy is probably the increase in temporal and spatial scale on which the economy can act. For example, no number of individual miners hacking away with picks and shovels can act on the scale at which modern mining equipment can act. The impact of modern mining techniques is qualitatively rather than just quantitatively different from the impact of the old techniques. This is virtually paradigmatic of self-organizational change. From the point of view of orthodox theory, industrialization and the development of technology can be made to look gradual and incremental, hiding the qualitative transitions. But then, of course, in some dimensions the differences between humans and chimps are exceedingly small as well and can be made to look gradual and incremental. Yet, the temporal and spatial scales at which humans operate on the environment are qualitatively different from those of the chimp.

MAKING WAVES

The most common self-organizing stuctures are cycles, including economic cycles. They remind us that structure can be temporal as well as spatial. The cycles—slow oscillations—that underly everyday life are obvious. Daily, weekly, monthly, and annual periodicities organize our lives. But cycles are ubiquitous at every known scale, from the vibrations of subatomic particles to the life cycle of galaxies. We ourselves function the way we do on the basis of a dizzying array of biological oscillations coupled to the oscillations in our environment.

For example, the following has a good claim to be the description of the general case:

> One possibility for the initiation and termination of biological rhythms is that an underlying rhythm is continuously maintained but that the organism can tap into and out of the rhythm by changing a control parameter (Glass and Mackey 1988, p. 82).

Glass and Mackey go on to exhibit a number of experimentally verified rhythms that occur naturally (and ubiquitously) in organisms. The study of biological rhythms, from those of the cell to those of organisms and groups of organisms, is currently a hot area of research. The rhythms are as short as high frequency atomic vibrations and as long as life cycles. These latter are both the most familiar and, in a way, the most puzzling. They include the prime number (13, 17, etc.) year periodicities of the locusts, and they include our own life cycles.

We can imagine some biota having evolved in which individuals were essentially immortal. But there would have been extreme problems with this arrangement. The sources of the trouble are animals—heterotrophs—who have to eat other organisms in order to live. A world of immortals would have to live entirely off each others' waste—or others' excess production, as in grazing. This is an imaginable circumstance, but it's not the way things have happened here on earth. Once heterotrophs came on the scene, continued existence demanded the fairly constant renewal of the food supply. So reproduction, already endemic to the chemistry of life, gave rise to the periodicity of generations.

There are two other considerations about the periodicity of generations that will be useful for us. The first is that the earth itself is a giant oscillator, enforcing the periodicities of the day and the seasons on everything that lives here. Organisms of the nonequatorial climates require a means to deal with both winter and summer. Many species of both plants and animals have evolved to deal with them by getting their reproductive cycles entrained to the seasonal cycles. In the easiest case, for example, annual plants winter over as seeds. In the more extreme climates, plants and animals have evolved even more elaborate metabolic periodicities, allowing hibernation and other wintering-over methods.

The second consideration is the inevitable slide to thermodynamic equilibrium. An immortal organism would have to have a virtually infallible system of repair at all levels, from molecules to organs. This is an evolutionary long shot. The evolutionary process has instead produced a system of longevity of lineage rather than longevity of individuals, a better thermodynamic bet, another reason for the periodicity of the life cycle.

We should now have a sense of the ubiquity of oscillations and vibrations—that is, periodicity. We can now ask whether there are socioeconomic periodicities; if so, where they come from; and, finally, whether they are good or bad vibes. The answer, of course, is obvious. There are all sorts of social periodicities, many of them entrained to the day, the seasons, the year, and the lifetime. We can orient ourselves in the landscape of coupled oscillators by means of the following:

> Our problem now is to imagine and locate the correlations between the rhythms of material life and the other diverse fluctuations of human existence. For there is no single conjuncture: we must visualize a series of overlapping histories, developing simultaneously. It would be too simple, too perfect, if this complex truth could be reduced to the rhythms of one dominant pattern. How clear, in any case, is that pattern itself? It is impossible to define even the economic conjuncture as a single movement given once and for all, complete with laws and consequences. Francois Simiand himself recognized at least two, when he spoke of the separate movements of the tide and the waves. But reality is not as simple as this relatively simple

image. In the web of vibrations which makes up the economic world, the expert can without difficulty isolate tens, dozens of movements, distinguished by their length in time: the secular trend, "longest of the long-term movements"; medium-term trends—the fifty year Kondratieff cycle, the double or hypercycle, the intercycle; and the short-term fluctuations—inter-decennial movements and seasonal shifts (Braudel 1973, p. 892).

Braudel's statement of the problem sounds like a straightforward invitation to the nonlinear dynamicist. This, in fact, is typical of the organizing framework of what is known loosely as the *Annales* school of history (Dyke 1990). Their basic focus on the behavior of long-term time series has produced a library of work just waiting to be linked to nonlinear dynamical models. Waiting beside them are people who have taken some relatively standard models from orthodox economics and examined their behavior in nonlinear regimes, or have taken nonlinear models from population biology (Lotka-Volterra models or simply the logistic growth equation) and shown how they can be applied to economic processes (Silverberg, as cited in Goodwin, Kruger, and Vercelli 1985; Gabish and Lorenz 1987).

We certainly shouldn't be surprised at the prevalence of economic cycles coupled to natural cycles. First, until very modern times, the cycle of the seasons was the dominant oscillator. Very simply, agriculture was seasonal. This made storage and preservation absolutely essential for people who were settled down anywhere. In the temperate zone, there was a long period when food couldn't be grown. In the tropics, spoilage was (and is) very rapid. One of the consequences of this during important periods in the world's economic history was that trade in salt and spices (for food preservation) dominated long-distance trade in general. This introduced another seasonal periodicity. In those days, long-distance trade took a lot of time and was possible (especially by sea) only in times of clement weather. On top of this, the naval and military activity necessary to secure this trade was itself coupled to the climatic oscillator. One of the immediate consequences of this was that the princely employers of military manpower would hire armies and navies in the spring, and fire them in the fall, pulling manpower out of the agricultural economy, then dumping it back in precisely when there was no need for it. This, in turn, produced chronic social unrest, affecting agricultural production itself.

Similarly coupled to the climatic oscillator were the availability of capital and its pattern of circulation. Money had to be raised to outfit the military and to capitalize the trading. Then, periodically and in rhythm with the seasons, capital would become available again as the ships came home. The typical economic pattern in these times was a feast or famine tidal slosh. The most conspicuous period of capital fluctuation occurred when the European economy came to depend on the influx of precious metals from the New World—also seasonal. Superimposed on all this was the quasi-periodic pattern of good agricultural years and bad ones.

In modern times, largely due to advances in transportation and the increased dominance of industry over agriculture, the seasonal periodicities have been damped. They still have their effect (they can still be the ruin of the small farmer, for example), but they have been considerably buffered. Nonetheless, economic cycles of various periodicities and quasi-periodicities continue to be identified (and experienced). So research into their origins retains its practical side.

A main line of research ties itself to theories of economic growth. The reasoning is, of course, that modern economies are characterized by a dynamics of growth and expansion. The equations favored by economists to model this growth turn out to be formally equivalent to equations for population growth commonly used in ecology. These equations are known to yield periodic solutions in some parameter ranges, and chaotic solutions in still other ranges. Thus, it is at least plausible to think that the same thing happens in economies. In particular, there are likely candidates for "forcing parameters" ready at hand that look as if they introduce vibrations into the system that could well be amplified into oscillations of various periods, organizing the economy into a wave-like motion over long periods of time. One such candidate is technological innovation, new production methods that start out small but which, if amplified above a threshold, go on to replace earlier methods. When models of this are investigated, they indeed turn out to yield periodicities, largely because of the sloshing of capital and labor that goes on at key stages of the transition (see Dyke 1994b).

However, even the standard models relating growth and investment, without any special attention to technological innovation, yield chaotic and periodic solutions under plausible parameter values (see Gabish and Lorenz 1987, esp. ch. 6) Investment seems to function as a forcing parameter very likely to amplify swings of an ongoing oscillator. It has also been noticed that the political systems of the modern world, especially the Western democracies and most especially the American democracy, have introduced periodicities of economic policy into the game. In the United States, these are most conspicuous in changes of presidency, but they also occur as a result of congressional elections. If stability is what we want, then there is a strong argument for electing congressmen every three years and having a presidential election every seven years rather than every four. The reasoning exactly parallels that used to explain the life cycle of the locust.

To sum up the situation, it's possible to say that even the work done so far has shown some basic and important things. At one time, periodic behavior of the economy was fully expected on the basis of long experience but surprising on the basis of the dominant theories. These theories were all grounded in the paradigms of equilibrium and self-regulation. But now, the use of new tools has allowed theorists to match up theoretical expectations with those grounded on experience. From the point of view of nonlinear dynamics, fluctuations and periodicities are the norm, and "relief" from them ought to be extremely difficult. And, of course, that is the experience of those who try to grapple with the economy in our time.

The situation, then, is much as it was in evolutionary biology in the middle of the nineteenth century. Fixity of species was then taken to be the norm, and how species would ever change was the mystery. But then Darwin, Wallace, and others set up the foundations of a theory of evolutionary dynamics, and *now* what seems surprising, intricate, and mysterious is how the biota of the earth manage to remain as stable as they do. Further, just as in economics, biological theory went through a long stage of demanding theories focused on equilibrium before it began to confront the dynamical situation head-on. In fact, the transition is far from complete at this moment, not so much among biologists themselves but among the ideologues of evolutionary theory. It goes almost without saying that a ground-level expectation of economic fluctuation, periodicity, and even chaos has very different policy implications than a ground-level expectation of equilibrium.

KNIGHTS OF THE HOLY BEAN

I began with a reminder of an inherited dualism, and I'll end with a few remarks about the same dualism, but now in the light of what nonlinearity, nonequilibrium, and complexity mean to us.

The choice—and it *is* a choice, not a necessity—to conceive the world in terms of a qualitative boundary between humans and everything else is perfectly understandable. I could hardly deny that, after going to the trouble of pointing out the qualitative difference between us and our nearest relatives, the chimps, in terms of the relative sizes of impact on the world. Furthermore, the industrialized nations have spent the last few centuries proving beyond all doubt that the attribution of these qualitative differences is well merited. If this weren't so, we would not be so worried about the environment. We shouldn't forget, however, that earthquakes and other natural "disasters" achieve a scale comparable to our own at their maximum. And galactogenesis occurs at a scale we will never reach. So, we are hardly the unique source of large scale processes. The fact remains that on scales and at rates characteristic of processes on earth, we've managed to become pretty powerful, though far from omnipotent. But the qualitative difference between us and everything else that we've now demonstrated is far from demonstrating the necessity and justice of the inherited dualism. For, inevitably, as our power has increased and been tested against natural parameters—and, plainly speaking, as we've learned more and more—our interconnections with everything else have been forcefully exhibited.

Setting ourselves off from everything else had its plausibility, at one time. So, correspondingly, it seemed to make perfectly good sense to throw ourselves into a scientific style that made our domination of "nature" an end in itself. For the institutions of science in the course of the last few hundred years, in fact, this goal of domination has been virtually essential, since science can only function when it has command of considerable social resources (money, of course, but also public confidence and public toleration), and science gained command of those resources by providing and/or promising the dominating technology.[8] So, Western science quite understandably grew into a comfortable self-conception centered on the ability to master nature. In a sense, it could think of itself as the fulfillment of the promise to Adam.[9]

The same story continues in the rise of the "social" or "human" sciences. They also learned to control social resources by promising to protect us from nature—including the "darker" parts of our own nature. Here, the successes have been far fewer than in the "natural" sciences, and we can see at the present time that this lack of success is quite reasonably leading to lack of public support.

Apparently exempt from this lack of success is economics. Modern economics began with advice on how to increase the wealth of nations and has, indeed, during its intellectual tenure, seen this wealth increase enormously. Economics, like the other sciences styled in conformity to the dualism, has succeeded in commanding resources from societies that themselves consider the nonhuman world as raw material to be exploited. The economic version of the dualism is that humans are thought of as rational economic agents, and everything else as commodity. The dualism lives in a metastable truce with the realization that human beings very often drift to the other pole of the dualism and become commodities themselves.

Earlier, I said that linearity seems the most natural thing in the world and that it conforms absolutely with the ideology of individual independence. Along an historical trajectory different from our own, *nonlinearity* might seem the most natural thing in the world. That would conform absolutely with an ideology of collaborative interaction—both between ourselves, and between ourselves and "nature." In fact, under that ideological view, it would be extremely hard to make the traditional dualism seem plausible—though, of course, we can't expect there to be any decisive ideology-free way to decide between ideologies.

Ideologies, like other social structures, change on the basis of complex social dynamics. Even if it were my deepest wish to "convert" people from one of these ideologies to the other, I would have no idea how to do it. The standard theories of how this is done are inadequate. Furthermore, the individualist, dualistic ideology is very deep and very well entrenched by now.

On the other hand, the sciences are, as it were, converted. For they've accommodated the new tools associated with nonlinearity and nonequilibrium as these tools have proved useful and relevant. The new-style experimental and analytic heuristics have followed along a bit more slowly. It turns out that the sciences born of the old ideology can find the continuities in their own work leading naturally to the adoption of the new style of science when the occasion arises. There are, of course, some less willing than others to make the changes, but that's no surprise.

Among the beneficiaries of the new ways of looking at complex systems are those interested in global climate models. Just in the last year or so, there has been a discernible shift in the ways these models are handled, and a shift in the characteristic expectations of what serious climatic change will look like. Probably because the limits of the capacity to predict are by now well understood in meteorology, working creatively with nonlinear models has been a relatively easy step for climatologists, and the unavailablility of precise predictions on the basis of climatological models is becoming an accepted way of life.

So all throughout the hard sciences, the new style of science is melding with the old. Most of the important introductory battles have been fought and won. This leaves us with the social sciences—in particular, economics. Their position is potentially interesting, and particularly that of economics. As the work of Samuelson and all the others shows, the mathematics of nonlinear nonequilibrium systems is well within the capacities of the mathematical economists. But the discipline of economics is sufficiently fragmented that this capacity may not have a substantial impact very quickly.[10] But second, the standard foundation of orthodox economics, the theory of equilibrium market price, rests on the assumption of linear systems near equilibrium as the paradigm standard cases. Giving up that assumption would be for the economists a far more radical departure than anything the physicists have had to confront. Physicists got a long way with two-body problems and experimental systems prepared to operate in a linear range. Economic systems are three-body systems virtually immediately and, as I showed earlier, the linear approximations to these systems can't be expected to be robust explanatory approximations. Physics was able to *extend* itself quite naturally into nonlinearity. Economics will have to be rebuilt (and the textbooks rewritten) to make the equivalent step.

Third, economists have long since learned the secret of being advisors to kings. Forecasting is one of their most publicly visible signs of expertise. A deliberate move into theory that has to give up the illusion of precise forecasting might be more than many economists can bear. We can't forget that, in contrast to most of the academic fields under discussion here, only a minority of economists are academics.

And fourth, economics was born in the same act of gestation as the individualist ideology. Economists deal with interactions grudgingly if at all, despite the fact that their subject matter is through and through a matter of interactions. The move to a style of their own science that actively confronts interactive self-organizing systems would for them be a sea change that may be more fundamental than we can expect them to handle. In this regard, we have to pay close attention to those economists who are already embarked on such a change and see how they fare in the discipline as a whole over the next few years.

One final way to look at the prospects for an improved economic science based on the study of nonlinear systems far from equilibrium is simply to summarize the points I've been making as they bear on ecological issues. Highly interactive systems can be expected to exhibit complex behavior, behavior whose patterns we can learn to understand even though detailed prediction can be expected to fail. The relationships between human beings, and them and everything else, are just such highly interactive ones. Attempts to simplify this complexity in terms of the summation of individual accumulation of what humans think they want and need in some brief period are bound to fail. These shortsighted accumulation strategies have produced what from a narrow and transitory point of view looks like success—perhaps measured in the awesomely ignorant statistics of "per capita income." The apparent success is an illusion, with the patent failure stored up in mountains of trash, unmanageable concentrations of ionizing radiation, mass extinctions, deterioration of habitat and diversity, and ever more frequent mutual slaughter among dense human populations shoved together like rats in a cage.

Here at the end of a millenium, the failures are getting hard to ignore. Indeed, the failures are beginning to dominate, rather than simply accompany, the successes. Entering the next millenium with a blind committment to the fundamentally mistaken simplistic intellectual strategies inherited from the European Enlightenment, linear orthodox economics can be expected to perpetuate and accelerate the failures—at our own expense.

Put this way, the point may feel peremptory, aprioristic, even simplistic. It's not. It takes what we all can plainly see as data, and what we have lately learned about the complexity of the world around us as object lesson. The web of relations constituting human life is not a two-body problem. We are not at or near equilibrium. The intellectual strategies marginally adequate to explain a session of baseball card trading are grossly inadequate to guide intelligent human life. They will produce no advances in human ecology. An improved economic science based on the acceptance of the heuristics of nonlinear far-from-equilibrium systems can not be just a prospect. It has to be an imperative.

NOTES

1. But Samuelson (1990) shows that neat maneuvres can be performed on interval rankings to make them behave well enough to support the dynamical treatment desired. He, in effect, shows how to produce a pseudo-natural scale for them. We will have several occasions to refer to Samuelson's elegeant little *tour de force*.

2. The "in general" is extremely important here. Boundary conditions can be imposed allowing local additivity of a limited sort. This isn't an unusual situation. The same is true of entropies (see Dyke 1992) and information (see Dyke 1994a). Local additivity, under the proper boundary conditions, is indispensible. Global additivity is an invitation to disaster.

3. For example, the symmetry assumption Samuelson (1990) employs to establish a common scale from preference data is, in fact, an assumption about additivity near equilibrium—as usual.

4. But also why (see Samuelson 1990 again), *given the first two laws of thermodynamics*, we can fix a natural scale for temperatures. Do similar laws exist for utilities?

5. Chaotic systems are ergodic *in the attractor*, not in the space, or representation, in which the attractor is embedded. This is why the statistics (e.g., Kolmogorov entropies) are so nonstandard.

6. The examples of linear nonequilibrium systems underlying the following discussion come from Caplan and Essig (1983).

7. One version of this system is called the Bernoulli map, and another is called the circle map (see, e.g., Jackson 1991, pp. 197ff).

8. And also providing the comfortable *feeling* that our scientific knowledge makes us masters even of phenomena that we can't actually control. This is why earthquakes and weather are so disturbing to those who cast their lot with totalized scientific dominance.

9. It's interesting to read Bruno Latour's *We Have Never Been Modern* (1993) in this light. Latour lays out the intellectual projects characteristic of modern times in the West in terms of the dualism we have considered, and attendant strategies of purification and translation designed to reinforce that dualism.

10. I once eavesdropped (out of boredom) on an introductory economics course of the standard sort. In response to a very good question from a student, the professor "replied," "Well, I don't think that we need to go into the higher mathematics of the situation." The higher mathematics he was referring to was algebra.

REFERENCES

Arrow, K. 1951. *Social Choice and Individual Values*. New York: Wiley.

Braudel, F. 1973. *The Mediterranean and the Mediterranean World in the Age of Philip II*. New York: Harper and Row.

Caplan, S.R., and A. Essig. 1983. *Bioenergetics and Linear Nonequilibrium Thermodynamics: The Steady State*. Cambridge, MA: Harvard University Press.

Dyke, C. 1983. "The Problem of Interpretation in Economics." *Ratio* 25(1): 15-29.

_____. 1990. "Strange Attraction; Curious Liaison: Clio meets Chaos." *Philosophy Forum* 20(4): 1-24.

_____. 1992. "From Entropy to Economy: A Thorny Path," Pp. 149-176 in *Advances in Human Ecology*, Vol. 1, edited by L. Freese. Greenwich, CT: JAI Press.

_____. 1994a. "The Mosaics of Monreale and Two-bit Theories of Information." *Bioscene* 220(2): 3-15.

_____. 1994b. "The World Around Us and How We Make It." Pp. 1-22 in *Advances in Human Ecology*, Vol. 3, edited by L. Freese. Greenwich, CT: JAI Press.

Gabish, G., and H-W. Lorenz 1987. *Business Cycle Theory: A Survey of Methods and Concepts*. New York: Springer Verlag.

Glass, L., and M.C Mackey 1988. *From Clocks to Chaos*. Princeton, NJ: Princeton University Press.

Goodwin, R.M, M. Kruger, and A. Vercelli 1985. *Nonlinear Models of Fluctuating Growth*. New York: Springer Verlag.

Jackson, E.A. 1991. *Perspectives of Nonlinear Dynamics*. Cambridge, MA: Cambridge University Press.

Latour, B. 1993. *We Have Never Been Modern*. Cambridge, MA: Harvard University Press.

Samuelson, P. 1990. "Deterministic Chaos in Axiomatic Utility Theory." In *Nonlinear and Multisectorial Macrodynamics,* edited by K. Velupillai. New York: New York University Press.

Swenson, R. 1989. "Emergent Attractors and the Law of Maximum Entropy Production: Foundations to a Theory of General Evolution." *Systems Research* 6(3): 187-197.

Velupillai, K. 1990. *Nonlinear and Multisectorial Macrodynamics.* New York: New York
    University Press.
Young, G.L. 1996. "Interaction as a Concept Basic to Human Ecology: An Exploration and
    Synthesis." Pp. 157-211 in *Advances in Human Ecology,* Vol. 5, edited by L. Freese.
    Greenwich, CT: JAI Press.

LINKING TECHNOLOGY, NATURAL RESOURCES, AND THE SOCIOECONOMIC STRUCTURE OF HUMAN SOCIETY:
A THEORETICAL MODEL

Mario Giampietro

ABSTRACT

This paper presents a model that describes socioeconomic systems as complex, adaptive, dissipative systems stabilized by informed autocatalytic cycles. The model can be used to analyze pre-industrial and post-industrial societies in their interactions with environments. Central to the model is the energy budget of society, which is approached as a dynamic equilibrium between the supply and requirement of useful energy per unit of working time. The supply of useful energy is determined by the amount of working time in society and its energetic return; the requirement is defined by the population size and the per capita energy consumption. A set of equations

Advances in Human Ecology, Volume 6, pages 75-130.
ISBN: 0-7623-0257-7

for dynamic equilibrium is described for individuals, the socioeconomic system as a whole, and the environment within which society operates.

Because of the complex nature of socioeconomic systems, two distinct perspectives have to be considered when analyzing societal development. One is the quasi-steady-state perspective, which analyzes the system at a particular point in time and space. It provides a definition and assessment of improvements in the "efficiency" of the system in terms of a given set of system goals and present boundary conditions. The other perspective is evolutionary, which allows for a definition and assessment of "adaptability" of the system, that is, its ability to perform well according to unknown future boundary conditions and different goals. These two contrasting perspectives cannot be blended into a single description of the system. Technological changes imply a trade-off between efficiency and adaptability, and never represent "absolute improvements." Biophysical analyses can provide indicators referring to both perspectives to analyze the nature and the effects of trade-offs but cannot provide a single denominator to define costs and benefits.

INTRODUCTION

Technological progress of human society has been analyzed by many authors in terms of an increased control of society over energy flows (see, e.g., White 1943, 1959; Cottrell 1955; Odum 1971; Pimentel and Pimentel 1979; Slesser 1978; Hall et al. 1986; Adams 1988; Tainter 1988; Gever et al. 1991; Debeir et al. 1991; Smil 1991; Olsen 1993a, 1993b). Besides energy, another biophysical parameter has been proposed, especially by sociologists, to study changes in socioeconomic systems, namely, human time allocation. This parameter enables an analysis of different roles and quantities of human time required for the various activities in socioeconomic systems (for a review, see Bailey 1990). The quantification of human time as an input in the socioeconomic process (both in consumption and in production) has been used by, among others, Zipf (1941) and Carlstein (1982). Their approach is slightly different from that developed in the field of classic economics, where the focus is on human labor as one of the major production factors in the economic process.

In this paper, I propose a combined use of the two biophysical parameters by assessing the patterns of allocation of energy flows and human time in the production and consumption processes in society. In particular, I analyze the development of socioeconomic systems by examining, at the level of society, changes in the absolute and relative consumption of useful energy in various socioeconomic sectors and changes in the absolute and relative amount of human time allocated to various activities.

This analysis, conceived as part of the process of self-organization of human society, is developed within a larger framework of analysis pertaining to the evolution of complex systems. That is to say, I approach the socioeconomic system as a *dissipative, hierarchical,* and *information-processing* system. The adjective "dissipative" implies that structures and functions of the socioeconomic system are stabilized by flows of matter that are sustained by a continuous process of energy dissipation. The adjective "hierarchical" indicates that the socioeconomic system is a holarchy, or a system made of "holons," where holons are defined as entities made up of smaller parts that simultaneously are parts of larger entities. The adjective "information-processing" indicates that with the use of certain "codes," it is possible for socioeconomic systems to store experience about what happened in the past and run simulations of what could happen in the future (computational capability and anticipatory systems). More details on these terms and references are given below.

Socioeconomic systems operate on multiple spatiotemporal scales. The scales include physiological processes within individuals, social relations within and among communities, micro- and macroeconomic relations, and biophysical exchanges of matter and energy flows between the society and the natural systems embedding it. As a consequence, socioeconomic systems can not be exhaustively described by a single scientific discipline that typically adopts only one particular window of observation at the time. This fundamental shortcoming of traditional, reductionistic science is the principal reason that I choose to build on complex system theory. Considering socioeconomic systems as complex provides a powerful tool for analysis and a wide array of possible applications. However, this choice requires a minimum of theoretical background. The next section is aimed at fulfilling this need.

Before plunging into complex system theory or getting into the details of the model, I first want to present the reader with a taste of my approach by presenting a simple example. Figure 1 depicts the population structures of four societies at different stages of socioeconomic development: (1) the Yanomamö (indios living in the tropical forest of South America), a society in which the population is stabilized by high mortality and fertility rates; (2) Burundi, a country in the first stage of the demographic transition; (3) the United States, a society close to the completion of the demographic transition; and (4) Sweden, a society that has completed the

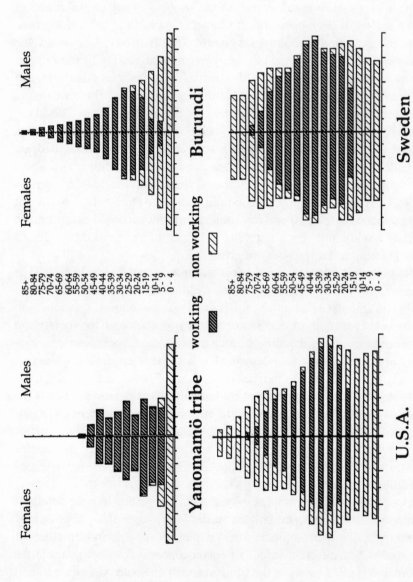

Figure 1. Age structure of societies at different levels of socioeconomic development.

78

demographic transition and whose population is stabilized at low mortality and fertility rates.

In these four societies, the ratio between the economically active and nonactive population ("working" and "non-working" in Figure 1) is markedly different. The ratios are closely related to the profile of human time allocation between labor and non-labor activities. In fact, the allocation of time to labor and non-labor activities is affected by the fraction of the population that is economically active, and their work load (number of labor hours per year). In turn, these two parameters are determined by several factors that have social significance, such as the distribution of the population among age classes (which is affected by life span), the minimum ages at which it is socially acceptable to enter and leave the labor force, and the fraction of the potential labor force that does not work for whatever reason (e.g., unemployment, education, illness).

The main idea behind the model is that the technological development of a society can be described in terms of an acceleration of energy throughput in the primary sectors of its economy, which then enables a change in the profile of human time allocation to different socioeconomic activities. In fact, a comparison between developed and less developed countries shows that technological development implies an increase in average per capita energy consumption—from less than 5 GJ/year[1] in developing countries to more than 300 GJ/year in the United States—and a dramatic reduction in the fraction of total human time allocated to work in the primary sectors of the economy—from 0.1 in developing countries to 0.04 in developed countries (Pastore et al. 1996). The latter reduction is due to a longer duration of the education process, a progressive graying of the population, a lighter work load for the working force, and an increase in the fraction of the working force (up to 60%) that is absorbed by the service sector of the economy. These changes in socioeconomic characteristics of society imply that, in modern societies, a smaller and smaller fraction of total human time is used to run the primary sectors of the economy (e.g., food security, energy and mining, manufacturing), while at the same time the material throughput in these sectors is dramatically increased. For instance, the Western standard of living is based on an energy throughput of more than 500 MJ per hour of labor in the primary sectors of the economy, while in subsistence societies it is lower than 10 MJ per hour.

The validity of the rationale that patterns of energy and time allocation are related to societal development has been checked by correlating the parameter "total energy throughput of a society per hour of labor in the primary sectors of its economy" to 24 classic indicators of physiological, economic, and social development for a sample of 107 countries representing more than 90 percent of the world population (Pastore et al. 1996). All these indicators of standards of living showed a good correlation with the parameter (see Figures 2 and 3). These findings confirm the existence of a direct link between profiles of energy and human time allocation and material standard of living.

The dramatic difference in the ratio of working time in primary sectors to total human time between developed and less developed societies is closely related to the trajectory of societal development understood as a complex system. Indeed, the difference can be explained in terms of a better control over energy and information flows in developed societies. This enables a switch from investments of human time in activities with a short-term return (steady-state efficiency) to investments of human time in activities with a long-term return (evolutionary adaptability). In short, technological development implies a reduced fraction of human time allocated to labor in general, but with a marked increase in labor activities that have a net consumption of energy—notably, services. The combined effect of these two trends is a dramatic decrease in the fraction of human time allocated to the primary sectors of the economy (or of human control allocated to stabilize the steady state). Of course, this result is possible only if a huge increase in labor productivity is realized in the primary economic sectors. Thus, the primary economic sectors must be capable of generating huge surpluses of energy and matter while absorbing only a negligible fraction of human labor time. In evolutionary terms, this means reducing the allocation of human time to repetitive activities (by reducing the redundancy of controls on the established set of activities performed in primary sectors) and increasing the allocation of human time on creative activities (by expanding the set of possible activities within society that translates into better adaptability).

In what follows, I first explore in detail the principles that can provide explanations, in terms of the evolution of complex systems, for the trends described above. Then, using a set of equations, I examine the nature of the biophysical constraints that affect the sustainability of different types of societies according to the process of self-organization.

In a second paper, immediately following (Giampietro, Bukkens, and Pimentel 1997, this volume), several applications of the model to preindustrial and post-industrial societies are provided.

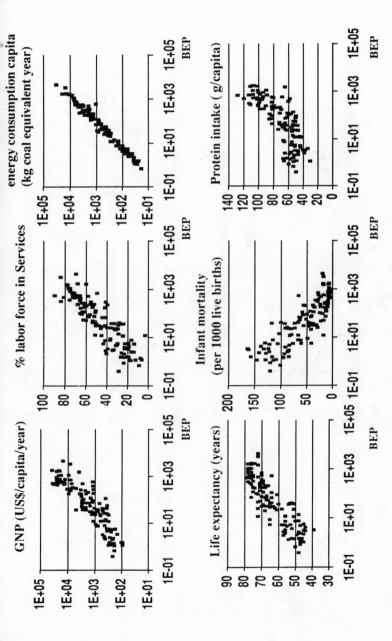

Figure 2. Correlation between social energy throughput per hour of labor in the primary economic sectors (BEP) and classic indicators of economic development.

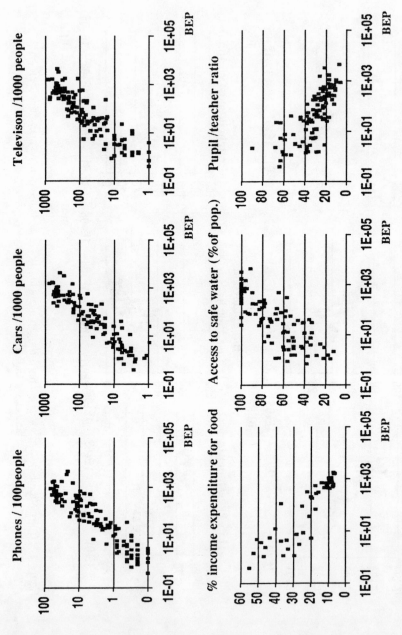

Figure 3. Correlation between societal energy throughput per hour of labor in the primary economic sectors (BEP) and classic indicators of material standard of living.

SOCIETY AS A COMPLEX SYSTEM

Dual Nature of Complex Hierarchical Systems

A system is hierarchical when it operates on multiple spatiotemporal scales—that is, when different process rates are found in the system (O'Neill 1989). In other words, systems are hierarchical when they are analyzable into successive sets of subsystems (Simon 1962, p. 468) or when alternative methods of description exist for the same system (Whyte et al. 1969). Human societies and ecosystems are perfect examples of complex hierarchical systems, and this represents a major complication in their analysis, especially when dealing with the issue of sustainability (Giampietro 1994a, 1994b).

Each component of a hierarchy may be called a "holon," a term introduced by Koestler (1969), which has a double nature (Allen and Starr 1982, pp. 8-16). A holon is a whole made of smaller parts (e.g., a human being is made of organs, tissues, cells, molecules, etc.) and at the same time forms a part of some greater whole (an individual human being is part of a family, a community, a country, the global economy, the biosphere). Therefore, hierarchical systems have an implicit duality: Holons have their own composite structure at the focal level, but, because of their interaction with the rest of the hierarchy, they perform functions that contribute to so-called "emergent properties" that can only be seen from higher levels of analysis. The problem in dealing with these entities is that the space-time closure of their structure (a lower-level perception) does not coincide with the space-time closure of their role (a higher-level perception). For example, Bill Clinton, the lower-level structure who is today the incumbent in the role of President of the United States, has a time closure within this role of a maximum of eight years, whereas the U.S. Presidency, as a function, has a time horizon in the order of centuries.

Because of their peculiar functioning on parallel scales, hierarchical systems can be studied either in terms of structures or in terms of relational functions. Established scientific disciplines rarely acknowledge that this unaviodable and prior choice of perspective implies a bias in the description of complex system behavior (Giampietro 1994a). For example, analyzing complex systems in terms of structures implicitly assumes a set of given initiating conditions (a history of the system that affects its present behavior) and a stable higher level at which functions are defined for these structures in order to make them meaningful and stable in time. Similarly, to have functions at a certain level, one needs to assume stability at the lower levels where the structural support is provided for the functions (Simon 1962). Hence, no description of the dynamics of a focus level, such as society as a whole, can escape the issue of structural constraints (*how—what is going on at lower levels?*) or functional constraints (*why—what is going on at the higher level?*).

Because of the duality inherent in any hierarchical system, the adjective "complex" is difficult to define. For instance, on the structural side, the complexity of a system has to do with the number of interacting subunits and the nature of their connections; on the functional side, complexity has to do with computational capability. Nicolis (1986, p. 7) equates an increase in complexity to an increase in the "length of the (most) compressed algorithm from which one can retrieve the full behavior of the system" or, in other words, to an increase in the amount and quality of information needed to describe its functions. Note that this definition implies the existence of a higher level at which something or someone is able to describe the system under analysis as having a particular function. However, the existence of this upper level (the observer as separated from the observed) introduces a series of epistemological problems (Barham 1990; Funtowicz and Ravetz 1990, 1994). The "full behavior" of a complex system not only cannot be retrieved but also it cannot be defined in scientific terms, because any description of a hierarchical system depends on an arbitrary choice of scale (what is the system?) and perspective (what is the system for?) (Giampietro 1994a).

Complex adaptive systems, in order to evolve, must enter into "communicative transactions" with their environments (Nicolis 1986; Salthe 1993). Again, this implies a hierarchical structure, because the systems need "one hierarchical level to perform cross-relation, and another, higher one in order to 'cognize' it" (Nicolis 1986, p. 2). (For more details on hierarchy theory, see Simon 1962; Whyte et al. 1969; Pattee 1973; Allen and Starr 1982; Salthe 1985, 1993; O'Neill et al. 1986.)

Complex Systems as Dissipative Systems: Efficiency versus Adaptability

As pointed out by H.T. Odum (1983, 1996), in economic and ecological systems that are based on a dynamic equilibrium of energy flows, an energetic investment (an activity), in order to be stable in time, must pay back its cost. In other words, an activity (a structure performing a function) must be able to take advantage of favorable boundary conditions to sustain the process of energy dissipation required for its own survival. On the other hand, the process of *exergy*[2] degradation needed to guarantee the metabolism of a dissipative structure has the effect of destroying the very same gradients of free energy on which the dissipative process depends. An exploration of the relation between structure and functions in the process of self-organization inevitably brings us to thermodynamic principles or, better, to principles of the thermodynamics of non-equilibrium. "A living thing is a semi-autonomous, temporal structure which preserves its organization in the face of the Second Law of Thermodynamics for a certain period of time by processing matter and energy flowing through the system in a coordinate way....as such, it belongs to the class of nonlinear dynamical systems away from thermodynamic equilibrium called *dissipative structures*.... Morowitz (1979) has shown that an energy flow through a system at steady state and away from thermodynamic equilibrium necessarily creates material flow cycles within the system" (Barham 1990, p. 200).

The following adds *how* living systems are able to accomplish this: Biological systems are complex, thermodynamic systems stabilized far from thermodynamic equilibrium by a process of self-organization induced by informed autocatalytic cycles (Brooks et al. 1989). Self-organization, therefore, means establishing a system of controls on matter cycles in order to regulate a process of *exergy* degradation in a way that stabilizes the system of control. Rosen (1991) has proposed this process, described as the establishment of a reciprocal entailment between two systems of entailments, as the basis of self-organization for living systems. In order to be sustainable in time, this autocatalytic process must be able to increase its rate of exergy degradation (larger and faster matter cycling) in time. This relates to two functions in the evolution of the system (Schneider and Kay 1994). The first is sustaining the short-term stability of the process by taking advantage of existing favorable gradients—that is, *efficiency* according to present boundary conditions (Conrad 1983). The second is sustaining the long-term stability of such a process by maintaining high compatibility in the face of a changing environment—that is, *adaptability*, defined as the ability to be efficient according to unknown future boundary conditions (Conrad 1983).

Recent recognition of these general principles has generated many new concepts and ideas in the field of evolution (see Ho and Saunders 1984; Depew and Weber 1985, 1994; Ulanowicz 1986; Wicken 1987; Brooks and Wiley 1988; Weber et al. 1988, 1989; Brooks et al. 1989; Layzer 1988, 1990; Barham 1990; Allen and Hoekstra 1992; Murphy and O'Neil 1994).

Stability of Complex Adaptive Systems as Resonance
Between Recipes and Processes

Prigogine (1978) observes that living systems are capable of establishing a resonance between coded information (e.g., DNA) that induces physical processes (e.g., metabolism) and physical processes that generate coded information. Simon (1962, pp. 477-482) states the same concept as follows: The stability of a complex adaptive hierarchical structure is due to the resonance between *"recipes"* stored in the structure of higher levels and *"processes"* (matter and energy flows) obtained by the interaction of lower level components. Complex systems stabilized by such a resonance have proven to be extremely stable in spite of their structural fragility at lower levels. For example, DNA "recipes" of some 10,000 years ago are carried in quite fragile components, such as certain species of hummingbirds, but today they are much better preserved than metallic artifacts constructed by humans only some 1,000 years ago. Because of their dual nature, resonating systems are able to absorb stress as functions at the level of the hierarchical structure as a whole, rather than as structures at the level of individual, lower-level holons. Small-scale changes in boundary conditions (such as fire, flood, or mechanical stress) can seriously affect lower-level holons (e.g., killing individuals of a species) but are easily absorbed by biological systems (e.g., the species), which are operating on large scales (O'Neill et al. 1986). Rosen (1991) describes the same mechanism in terms of relational organization. Dissipative systems hierarchically organized on several spatiotemporal scales are able to make models of themselves by establishing a mechanism of reciprocal entailment between systems of entailments operating on different scales and, therefore, at different speeds (see also Rosen 1985).

A second important feature of resonating systems is that the *information system,* where coded information is stored and replicated, provides the possibility for generating *novelties* (for example, through errors in replication or sexual reproduction). This enables resonating systems to generate a variability needed to adjust to slow, large-scale changes in boundary conditions and to co-evolve with their environments (Salthe 1993). In higher-evolved systems, information processing is further boosted by the development of sophisticated "anticipatory systems" (Rosen 1985) able to run simulations of boundary conditions (e.g., thinking in humans).

The duality between structures (incumbents) and functional relations (roles) within hierarchical systems—when seen from different levels—is crucial also for the stability of human societies. For instance, a doctor in a hospital has a function (role) that depends on the societal organization rather than on the condition of the particular physician (incumbent) hired for the job. The position will remain in the hospital if the particular physician leaves. Only when a lower-level perturbation is amplified onto a scale large enough to affect the functioning of the entire system are lower-level processes (incumbents performing a particular role) able to force changes in the organization at the higher level. To stay with our example, this would be the case should all the physicians in the country agree to a labor strike. But in this case, the ability of lower-level holons to organize themselves on a larger spatiotemporal scale takes the form of an *emergent property*, that is, a new function detectable on the higher level: It is a sign of the existence of a national union of physicians.

The dual nature of complex systems and their components has important consequences for the analysis of socioeconomic systems. Here, I use the conceptual distinction proposed by Bailey (1990, p. 119) between *role*, or the function defined on the higher level, and *incumbent*, or the structure as seen on the lower level and performing the particular function on the higher level. Salthe (1993, p. 174) proposes the terms "types" and "individuals," respectively, for the same concepts.

The very same concepts of resonance between recipes and processes (reciprocal entailment between systems of entailments) that are used to describe biological evolution can be applied to the analysis of human societies. For example, Bailey (1990, p. 186) refers to this resonance this way: "The relationship between symbols and action is reciprocal and circular (and cyclical). That is, symbolized concepts are used to guide human interaction. This leads to a transformation of the synchronic symbol structure,...which leads to a transformation of concrete action in the next period, and so on." In other words, this passage says that structure at the higher levels stores "images/recipes" (synchronic symbols), and because of the hierarchical relation, such a structure affects actors (lower-level holons) and, therefore, processes. In turn, the ensemble of processes taking place at the lower level (when aggregated on a larger scale) will induce changes in the "images/recipes" (system of controls) of the higher level in the long term (providing an update of the set of symbols).

Theoretical Framework

Using the concepts and ideas of complex system theory discussed above, I shall now briefly sketch the theoretical framework within which the model is developed.

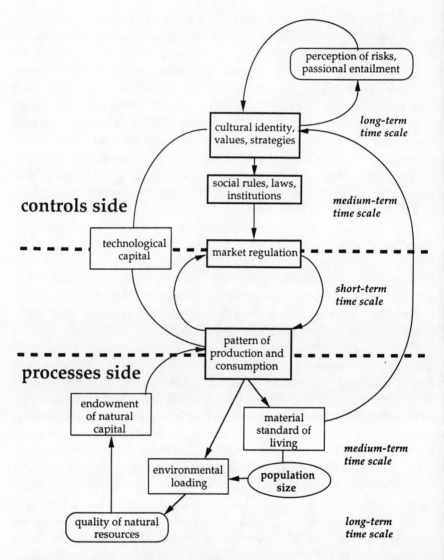

Figure 4. Resonance between controls and processes in socioeconomic systems.

Humans alter with their technology, the ecosystems in which they live in order to increase the efficacy of the process of production and consumption of goods and services in society. In other words, they attempt to stabilize and "improve" the structures and functions of society according to a set of internally generated values. The basic process of resonance that stabilizes structures and functions of socioeconomic systems is illustrated in Figure 4.

On the process side, self-organization can be seen as the ability to stabilize in time a network of matter and energy flows, such as the stabilization of the process of production and consumption of goods and services in society. The characteristics of a particular self-organization process depend on the control side, which represents experience from past events accumulated in a complex system of controls. For instance, the cultural identity of a nation (e.g., religion, mores, values), which is high in the hierarchy of controls, influences institutional and political settings, which in turn define the rules, laws, and regulative mechanisms of everyday activities. Thus, the control side has two important functions. First, it acts as a filter on what the economy can and can not do, for instance, by imposing laws that regulate what can and can not be produced and consumed. Second, it establishes criteria to redistribute the "disposable" power and income among the various social groups in society. In other words, it shapes the curve of the distribution of collective properties (parameter values defined at the societal level) over social groups and individual human beings.

On a different (larger) temporal scale, we find the reverse: The process side influences the characteristics of the control side. Indeed, external or internal changes on the process side, such as the discovery of new natural resource inputs, access to new technologies, or warfare, can change the perception of what is important or valuable for society. When such a change in the perception of values is stabilized for a period of time long enough to affect the system of controls in society, the direction of exerting control is reversed: Lower-level changes on the process side bring about changes in political institutions or laws through a change in the cultural identity of society.

Thus, putting it concisely, socioeconomic systems are *dissipative*—they tend to destroy the gradients on which they thrive; they are *holarchic*—they operate in parallel on several spatiotemporal scales; and they are *adaptive*—they tend to become, sooner or later, something else. These characteristics have two important implications.

First, looking at socioeconomic systems at a particular point in time and space (considering them in quasi-steady state), their current structure is based on a particular pattern of stabilization of the societal metabolism. That is, given initiating conditions, cultural values, goals, and strategies, we can check the stability of the current dynamic equilibrium (quasi-steady state) by checking the mutual compatibility between the structure/function of the socioeconomic system (internal constraints) and the structure/function of the ecosystems with which the socioeconomic system interacts in its process of self-organization (external constraints). Note, however, that the term *quasi*-steady state indicates that this equilibrium is based on the "ceteris paribus" hypothesis, or presumed stability, of boundary conditions and initiating conditions. This assumption can not be held when long periods of time are considered.

Second, when studying the evolutionary trajectory of socioeconomic systems, we need to consider variables that refer to "value judgment." What is the effect of "passional entailment" (personal feelings and individual aspirations in the determination of self-fulfilling prophecies) on possible paths of evolution of the socioeconomic system? What are the power relations between different social groups? How does cultural mode-locking (outdated sets of images about risks and opportunities stored in the system of controls) affect the choice between alternative paths of evolution?

When the two views on the process of self-organization of socioeconomic systems are considered (efficiency versus adaptability), it is clear that the effects of technological development have to be studied on several hierarchical levels (e.g., ecological, economic, and sociological terms) and according to several perspectives (e.g., individuals, families, social groups belonging to the society, the society as a whole, the ecosystem that embeds the socioeconomic system).

The simple matching of biophysical flows at a particular point in space and time assessed at a particular level (assessing compatibility with physical laws in terms of average values) is not sufficient to guarantee the stability of socioeconomic systems. A particular equilibrium between energy supply and energy requirement of societal metabolism must also be viable in socioeconomic and ecological terms. Viability in socioeconomic terms requires that collective properties of society are compatible with the structure and function of lower-level components (e.g., families, individuals). The process of self-organization of society must respect the integrity of lower-level holons that make up the society. This compatibility refers to both physiological processes, such as food requirements and limits to the physical activity (labor) of human beings, and to social processes, such as "acceptable" standards of living and issues of social equity (distribution over lower-level holons). Viability in ecological terms requires that the process of self-organization of society must respect the integrity of higher level holons, such as the ecosystems with which the society is interacting. This implies that human activity should preserve (or at least avoid the rapid degradation of) the stability of the boundary conditions on which societal metabolism depends.

BUILDING BLOCKS OF THE MODEL: CONCEPTS AND IDEAS

Dual Nature of the Informed Autocatalytic Loop: Basic Concepts

To remain away from thermodynamic equilibrium—an essential requisite to have any process of self-organization—complex systems have to stabilize in time the process of energy dissipation (or, better, *exergy* degradation; see note 2). The process of self-organization of society can be described in terms of two types of activities, those related to *efficiency* and to *adaptability*.

Activities related to efficiency are those that search to stabilize the current process of energy dissipation according to existing stored information about boundary conditions, in order to match in time and space the flows of materials required by societal metabolism. Hence, by increasing its efficiency, society tries to stabilize and strengthen its current pattern of interaction with its environment. In energetic jargon: In its everyday routine, society works to remain in the steady state that was individuated during its history, from among the possible processes of energy dissipation feasible according to physical laws, initiating conditions, and boundary conditions. Expressing the same concept in a more traditional way, an increase in efficiency means stabilizing the existing set of activities by making available to society a larger and more reliable flow of useful energy. The system of controls used by society to pursue this goal is based on the existing (old) set of images/recipes of boundary conditions, that is, on the picture generated and stored on the basis of past interactions of society with its environment.

Activities related to adaptability are those that expand the possible solutions that will stabilize the process of energy dissipation according to future changes in boundary conditions. These would be activities that are able to match in the future the (unknown) needs and aspirations of the socioeconomic process during its continuous evolution (in the long run) when facing different (unknown) boundary conditions. By improving its adaptability, the system increases the probability of reaching new situations of quasi-steady state in thermodynamic nonequilibrium in the future, when boundary conditions and internally generated goals will have changed. Thus, adaptability refers to the ability to find new ways of stabilizing processes of energy degradation that will keep alive the resonance between processes and recipes (the reciprocal entailment between two systems of entailments) in spite of a changing environment and a changing identity of the system.

The process of self-organization of society can, therefore, be analyzed two ways: (1) on the process side, by assuming the system as in a quasi-steady state and studying the pattern of investments of useful energy; and (2) on the control side, by studying the profile of human time allocation to efficiency and adaptability from an evolutionary perspective. On the process side, on a short time scale, the useful energy dissipated by society (level of energy degradation stabilized in time) can be assessed by the inputs of exosomatic energy consumed[3] by society and allocated to various activities. On the control side, on a longer time scale, the activity of controlling the societal process of self-organization can be assessed by the ratio of labor to non-labor time (e.g., in modern societies, the distribution of human time between producing and using added value) and the pattern of distribution of labor time among the various economic activities.

Thus, in effect, we have to perform a dual analysis of societies. The first is to analyze the process of energy degradation that stabilizes structures and functions of society as seen in the steady state (a snapshot picture). This implies analyzing the set of internal constraints, such as the processes of energy conversion used in societal metabolism, the main energy sources used (e.g., wind, biomass, fossil energy), and the characteristics of the devices generating power (e.g., machines, animals) that determine the energy costs of generating throughputs (e.g., input/output of processes); and the set of external constraints, such as the resilience of ecological processes supporting the society and the availability of resource inputs and sinks.

The second is to analyze the nature of the system of controls that regulates flows of matter and energy in society. This means looking at the process of self-organization in evolutionary terms. The sustainability of any solution will be affected by the ability to adapt to changes imposed on the dynamic energy budget, either by new boundary conditions (issuing from the nature and speed of changes in the ecosystem) or by internally generated perturbations (e.g., Malthusian instability or a drive toward a better material standard of living). This translates into the ability to quickly adjust the profile of power delivery in different activities (e.g., labor productivity in different economic sectors) and to encode, store, copy, transmit, and process information (e.g., size of the service sector and household consumption).

The model uses parameters that describe human societies at the level of the individual (e.g., average body weight of individuals, expected life span, metabolic flow) and parameters that refer to the focus level of society as a whole (e.g., energy consumption per capita, fraction of population that is working, fraction of the working force in the primary economic sectors). When society is characterized in terms of total demand of energy inputs or in terms of total matter throughputs, it is possible to put the process of self-organization of society in relation to the higher hierarchical level, that is, the ecosystem in which the society is operating, by using the concept of "environmental loading" (which compares the scale of human activity with the scale of ecosystem activity). Variables used in the model include both *intensive* (e.g., metabolic rate and labor time as a fraction of total human time available) and *extensive* variables (e.g., total population mass, total labor time available). In this paper, the variables are specified as they appear.

Looking from the Process Side: Society in Quasi-Steady State

Analysis of Energy Flows

Analyzing ecosystem structure, Ulanowicz (1986) finds that the network of matter and energy flows making up what we call an ecosystem can be divided in two parts. One part generates a hypercycle—that is, a part that is a net energy producer for the rest of the system. Since some dissipation is always "necessary to build and maintain structures at sub-compartment levels" (Ulanowicz 1986, p. 119), this part comprises activities that, taking advantage of sources of free energy outside the system (e.g., solar energy), generate a positive feedback by introducing degradable exergy into the system at a higher rate than it is consumed. The role of this part is to drive and keep the whole system away from thermodynamic equilibrium. The other part has a purely dissipative nature. This part contains activities that are net energy degraders. However, this second part is not useless for the system: It has the role of providing control over the entire process of energy degradation and stabilizing the whole system. An ecosystem made of a hypercyclic part alone could not be stable in time. Without the stabilizing effect of the dissipative part, a positive feedback "will be reflected upon itself without attenuation, and eventually the upward spiral will exceed any conceivable bounds" (p. 57). Hypercycles alone cannot survive; they just blow up.

A similar approach is used in the model to describe society from the process side: Society consists of two compartments, one of which is hypercy-

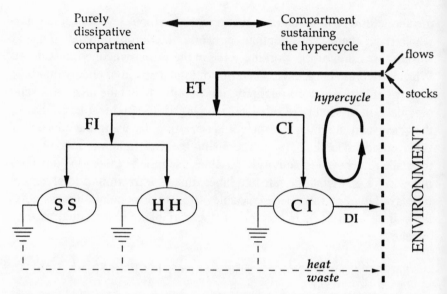

Figure 5. Socioeconomic structure seen from the process side: Profile of allocation of useful energy.

ET is the total energy throughput in society. FI and CI stand for fixed and circulating investment of useful energy, respectively. DI is the direct investment of useful energy in the interaction with the environment. HH is the household sector, including non-encoded activities with a net energy consumption. SS is the service sector including encoded activities with a net energy consumption. CI stands for the primary economic sectors that comprise all encoded activities with a net energy return.

clic (a net producer of useful energy for the rest of society) and the other purely dissipative (a net consumer of useful energy). This approach is illustrated in Figure 5, where the energy consumed by society is allocated to three sectors involving distinct types of activities:

1. Activities in the household sector (HH) are purely dissipative in nature as they consume net energy in the short term. They are not strictly encoded in the form of defined roles or protocols. These activities include sleeping, personal care, leisure time, and all activities performed by the economically inactive population.

2. Activities in the service sector (SS) are also dissipative in nature but are "encoded" in the form of defined social roles (e.g., job positions). These activities include all services such as police, army, health care, education, insurance, and so forth.

3. Activities in the primary economic sectors (CI) have a positive return in terms of energy flows and are "encoded" in the form of

defined social roles (e.g., job positions). This includes the energy and mining sector, the manufacturing sector[4] in modern economies, food security, and environmental security. CI activities generate the hypercycle by stabilizing the autocatalytic loops of endosomatic and exosomatic energy. They must be able to generate the energy surplus consumed by the dissipative sectors HH and SS.

The Role of Purely Dissipative (FI) and Hypercyclic (CI) Activities

Energy input (ET) is converted in society into useful energy for two main purposes. One is to run, replace, and maintain the endosomatic energy flow and the exosomatic energy supply sector (CI activities in Figure 5). This flow of useful energy (CI) is needed to stabilize the flow of energy input in time and can not be used for alternative purposes. The second purpose is to generate a "spare" flow of useful energy (FI in Figure 5) that humans can allocate to whatever other activities they judge valuable, such as health care, cultural activities, religion, and recreation. Thus, this includes the energy allocated to activities in the service sector (SS) and household sector (HH). The level at which the FI flow is stabilized can be assumed to be proportional to the material standard of living reached in society. In analogy with economic terms, this spare power (FI) can be considered the "disposable energy income" of society.

The different nature of the use of the two energy flows FI and CI can also be seen in terms of hierarchy theory. The energy used in the energy supply system (CI flow) is used to maintain the dynamic energy budget in the *short term*—on the time scale of operation of the energy converters (e.g., for exosomatic devices a life span of 10 years)—by feeding and replacing the exosomatic compartment and the humans providing the controls for it. The "spare" useful energy allocated to activities elsewhere (FI flow) will affect the dynamic equilibrium of society in the *long term*, for example, by accumulation of knowledge and capital and expansion of human potentialities. Thus, in analogy with economic terms, the *circulating investment* (CI flow) refers to energy spent directly in the primary sectors of the economy and with effects detectable on a short time scale, whereas the *fixed investment* (FI flow) refers to energy spent in the maintenance of the rest of society's activities with effects detectable only on a longer time scale.

It may be clear that the quantity of useful energy that any society can allocate to the stabilization of its structure in the long term (FI flow) depends on the efficiency of the energy supply system (ET/CI). Hence, we have here a biophysical constraint, given boundary conditions and given a defined performance of technology, on the fraction of useful energy that can be allocated, at the level of society, to "adaptability." A higher efficiency of the energy supply system is indicated by a lower demand for useful energy consumed for its own operation and maintenance per unit of energy throughput (by a higher ratio ET/CI).

Biophysical Constraints Related to the Energy Budget

The description of the energy budget of society, when read from the process side, requires only a few parameters (Figure 5):

1. *Total energy throughput (ET)*—that is, the total amount of energy "metabolized" by society. Hence, ET is the energy used for the stabilization of societal structures and functions under human control. Energy used by humans in the process of self-organization can be either *endosomatic energy,* that is, food energy converted within human bodies (physiological conversion) that belong to the social system, or *exosomatic energy,* which includes all other conversions of energy that are performed in society outside of the human bodies that belong to the society.[5] The latter include animal power, fire, wind power, engine power, electric power, and the power of human slaves.

    When individual humans are organized into a society, they manage to harness a flow of exosomatic energy that is greater than the sum of the endosomatic energy expended by all individuals comprising the society (Giampietro et al. 1993). Therefore, the ratio between exosomatic energy and endosomatic energy consumption (exo/endo ratio) is always higher than 1. Such a ratio can thus be used as an indicator of the multiplicative effect that knowledge and technology induce in human activity. This model describes the self-organization of human society as a process driven by an autocatalytic loop of useful energy (energy invested to get more energy). Since the final result of this process is the stabilization of the flows of endosomatic energy obtained through the stabilization of flows of exosomatic energy, *the energy throughput (ET) in this model is only exosomatic energy.* The flow of endosomatic energy is no longer considered as a relevant parameter for the stability of the energy budget, because changes occurring in the human compartment (flows of endosomatic energy) are already considered in terms of changes in the profile of allocation of human time.

Under this perspective, society is viewed as a sort of superorganism, made up of humans, with its own metabolic rate. The energy throughput (ET) of society is then simply the product of the body size of the superorganism (human mass assessed by the endosomatic flow) and its metabolic rate (the exo/endo ratio). At this point, the flow of endosomatic energy disappears in the equations. At the level of society, the "efficacy" of human activity is proportional to human time and the flow of exosomatic energy is controlled (ET), but no longer to flows of endosomatic energy.

2. *The direct investment of useful energy (DI)*—that is, applied power invested by society in the interaction with its environment in order to sustain societal metabolism (making matter and energy inputs available). Hence, DI is the fraction of useful energy invested by society to obtain and use energy resources and matter inputs. Again, (see footnote 3) the amount of useful energy spent in different activities is distinct from the amount of energy input that is consumed. Indeed, there is an inevitable loss in any energy conversion, as is illustrated by the dissipation sign on the left side of each compartment in Figure 5. For instance, only 20 percent of the energy content of gasoline is delivered in the form of applied power by a thermal engine. Similarly, only a small fraction of the energy content of total food consumed by a population is delivered in the form of applied muscle power in labor activities (Giampietro and Pimentel 1991a, 1992; Giampietro et al. 1993).

3. *The circulating investment (CI)*—that is, the fraction of total energy throughput allocated to the compartment sustaining the energy hypercycle (CI activities). This comprises the energy consumed in direct interaction with the environment (DI; in industrialized societies, this is energy consumed by the energy sector and in mining, food security, and environmental security) and, in an indirect way, in the construction and maintenance of exosomatic devices (in industrialized societies, this is the energy consumed by manufacturing).

4. *The fixed investment (FI)*—that is, the fraction of total energy throughput allocated to the compartment of pure energy dissipation (HH and SS activities). In modern societies, this would be the fraction of useful energy absorbed by the household sector and service sector.

A key relation between parameters is the ratio FI/ET. It indicates how much of the energy throughput consumed by society is "disposable" energy usable for building adaptability (see Figure 5). This is a fundamental parameter that measures the strength of the hypercycle that sustains the process of self-organization of society and, therefore, the "efficiency" of the exosomatic energy loop.

Clearly, since $ET = CI + FI$, we could also use the ratio CI/ET, where $CI/ET = 1 - FI/ET$. This ratio indicates how much energy is absorbed by the hypercycle for its own operation. Parameters that can improve the efficiency of the hypercycle are those that reduce the requirement of energy input for the circulating investment (CI/ET). Five parameters are considered in this model.

First, the efficiency of power generation (η), which is defined as the thermodynamic efficiency at which exosomatic devices convert energy input into a flow of useful energy (useful energy/energy input). It follows that the cost of generating useful energy is represented by ($1/\eta$).

Second, the energy spent in manufacturing exosomatic devices (B&M), which is the indirect cost of generating useful energy. The ratio (B&M)/ET indicates how much energy used by society is spent in building and maintaining exosomatic devices discounted over their life span.

Third, the quality of energy and mineral sources (E&M), which is the energy spent for making accessible the required input of energy and raw material to society. The ratio (E&M)/ET indicates how much energy used by society is invested in the energy sector. When the ratio (E&M)/ET is high, a large fraction of the useful energy used by society (and, therefore, of ET) must be invested in simply "procuring energy input" and, consequently, the diversity and complexity of activities within such a society will be limited, which is to imply the ratio FI/CI has to be small. The same applies to the energetic cost of other raw materials required by society, such as the quality of mineral ores.

Fourth, the energy spent to provide food security, which is considered equal to the useful energy spent in the food system (FS). This parameter can be approximated on the basis of the exo/endo ratio (exosomatic energy spent as input per unit of food energy output).

Last, the energy spent in providing environmental security (ES), which is approximated by the useful energy spent in monitoring, pollution control, and restoration of degraded ecosystems. ES depends on the scale of the economic process (population size and exo/endo ratio), compared to the scale of natural capital.

Environmental and Technical Constraints

In order to characterize the process of energy degradation, we need to specify the processes used to generate the useful energy required by the various activities in society. Depending on the nature of the energy input required for specific processes, *external constraints* (boundary conditions) will define the availability of energy and material inputs and the availability of waste-specific sinks. Depending on the nature of the conversion process of energy input into useful energy and type of metabolism, an available technology—an *internal constraint*—will define the threshold to which the activity of self-organization can be expanded.

Regarding the external constraints for different types of energy input sources, the distinction proposed by Georgescu-Roegen (1971, p. 226) between *stock* and *fund* services[6] is useful in the definition of external constraints. When the flow ET is generated from a *fund*-type energy resource (e.g., renewable biomass), the total annual energy demand implies the requirement of a certain amount of space to fix solar energy into biomass. In general, the larger the energy throughput per capita, the larger the area that must be exploited per capita and the lower the efficiency of the energy sector, as more work has to be performed to collect and transport the biomass from the larger area. This is the reason for the weak hypercycle typically found in pre-industrial societies, which is discussed in detail in Giampietro et al. (1997, this volume).

On the contrary, when ET is generated by *stock*-type energy sources (nonrenewable sources such as oil), ET can be increased virtually infinitely, depending mainly on the internal ability of society to make good use of the available energy input (e.g., if the supply of energy is linked to the supply of useful power to be invested in the energy sector) and depending on the boundary conditions for the matter cycles coupled to societal metabolism. Clearly, the larger the energy throughput ET, the faster will be the depletion of energy stocks and the more likely the pollution problems (because of the larger the stress on the stability of boundary conditions).

The Issue of Scale

Virtually all analyses of the biophysical structure of society in terms of energy flows assume the energy flows used by society to be in quasi-steady state (e.g., Odum 1971, 1996; Gilliland 1975; Slesser 1978; Jørgensen 1992). This means that, at a particular moment in time, boundary conditions and technical coefficients are assumed to be fixed. This assumption is reasonable, but it must be kept in mind that the particular time scale (e.g., short/medium versus long) chosen to describe the self-organization process of society affects the range of possible applications of the analysis. In particular, information describing the structure of energy flows in a particular society at a particular moment in time is not sufficient to deal with the problem of sustainability. In fact, sustainability involves dealing with changes in both the internal characteristics of society and the boundary conditions. This, in turn, requires an analysis of the self-organization process of society in relation to the individuals belonging to it, as well as a comparison of the size of the self-organization process of society (the demand of inputs and waste disposal) with that of the biophysical system containing society (the environment). Indeed, the stability of boundary conditions for human society is reflected by the relative size of the natural capital (the size of biophysical processes generating favorable situations with regard to stocks of resources and sinks for outputs) compared to human-made capital (the size of socioeconomic processes whose stabilization depends on degrading favorable biophysical conditions).

Looking from the Control Side: Society as an Evolving System

Profile of Human Time Allocation

In their social interaction, humans experience limits that "are related to the fact that a human being is more nearly a serial than a parallel information-processing system" (Simon 1962, p. 476). When the capacity of lower-level holons to interact is saturated, the system faces an internal constraint to its further expansion (e.g., because it takes time to care for a friend, a person cannot have an infinite number of them). Hence, the limited ability of humans to exert control over flows of useful energy allocated to various activities is a fundamental factor in shaping the process of self-organization of human society. Redundancy in the requirement of controls on the established set of activities needed to stabilize the current steady state reduces the ability to expand the set of activities performed by social systems.

It is useful here to apply some elements of ecosystem theory to the study of the evolution of human societies, and to this purpose I will briefly explore the analogy between ecosystems and human societies. First, we need to define an equivalent of "species," the holons of ecosystems, for the organization of human society. A reasonable candidate would be functional role in the economy, such as "labor position." In fact, a labor position in society, like a species in an ecosystem, indicates the ability to perform an encoded activity, that is, an activity that has proven to be useful for the system from past experience. However, labor positions alone are not sufficient to regulate flows of matter and energy in society. During labor time, humans control only flows of resources that are used in the economic process of production (the supply side). In ecosystems, it is well known that autotrophs (primary producers) need heterotrophs (herbivores, carnivores, and detritus feeders) to degrade their by-products and to recycle nutrients. Thus, the rate and pattern of primary production in ecosystems are controlled not only by primary producers but also by the activity of consumers and decomposers. Indeed, the more developed the ecosystem, the more consumers and decomposers play a key role in regulating the overall flow of solar energy. In order to produce more and better, plants must be "eaten" at a faster rate by heterotrophs.

The same applies to the economic process: In order to be able to produce more, society must consume more. As with heterotrophs in the ecosystem, the amount and pattern of human consumption in society directly affect the amount and pattern of production.

The direct biophysical relation between autotrophs and heterotrophs in ecosystems implied by the word eaten is intriguing when it comes to socioeconomic analyses. In order for society to be more productive and efficient, labor hours must be partially "eaten," that is, reduced in number, by becoming part of the consumer compartment. In other words, at the level of society, labor time has to be sacrificed in favor of consumption time if more products and services and higher wages are to be obtained. Here, the distinction made earlier between roles and incumbents is particularly useful: When society goes through a phase of economic development, it can change the allocation of human time among different roles (reducing the time in production and increasing time in consumption) with the same endowment of incumbents (same population structure) simply by a change in socioeconomic variables, such retirement age, length of education, and work load. Sooner or later, of course, one should also expect changes in the distribution of the population among age classes (see examples in Figure 1 and the discussion of the process of demographic transition in Giampietro et al. (1997, this volume).

"An input-output matrix of the economy...reveals the nearly decomposable structure of the system—with one qualification. There is a consumption subsystem of the economy that is linked strongly to variables in most of the other subsystems. Hence, we have to modify our notions of decomposability slightly to accommodate the special role of the consumption subsystem in our analysis of the dynamic behavior of the economy" (Simon 1962, p. 475). In other words, in order to boost the performance of an economic system, we not only need more diversity and more efficiency in labor positions (roles) but also more diverse and better consuming roles. As noted earlier, labor positions and consuming roles are defined at the hierarchical level of society (above that of individuals) because their existence is independent of the incumbent at a particular moment (Bailey 1990).

The possibility that shortage of human time in consumption can negatively affect the performance of an economy has been discussed by Zipf (1941): "Expressed differently, in 1929 the United States discovered a new 'raw material': *leisure time*, which in a way is just as much a 'raw material' as coal, oil, steel or anything else, because for many types of human activity, *leisure time* is an essential prerequisite...any change in kind or amount of goods or of processes within a social-economy will necessitate a restriation[7] within a social-economy itself" (p. 324).

Increasing the time allocated to leisure has an important effect on adaptability: Labor roles due to organization can be seen as replicated actions (Bailey 1990, p. 179) and, therefore, they reflect what has happened in the past. Also, leisure time tends to be allocated over an "established" set of leisure roles as individual choices are constrained by cultural identity. For instance, Europeans watch soccer whereas Americans prefer baseball or football. However, fidelity to leisure roles is less strictly enforced by society than that of labor roles. This allows more freedom of decision for individuals and, hence, more variability in the set of activities performed during leisure time.

Figure 6 depicts the profile of human time allocation among the three sets of activities introduced earlier in Figure 5. A corresponds to the household sector HH, B to the service sector SS, and C to the primary economic sectors CI. The fraction of human time allocated to the household sector (A) has to do with adaptability, because it is invested in building up cultural identity and stimulates innovations that may benefit future adjustments in the activities in society (long-term stability). Time allocated to C activities has to do with efficiency, since it concerns repetitive encoded activities based on experience accumulated in the past and, therefore, is considered "circulating investment" (CI). In terms of evolution, the role of activities in the ser-

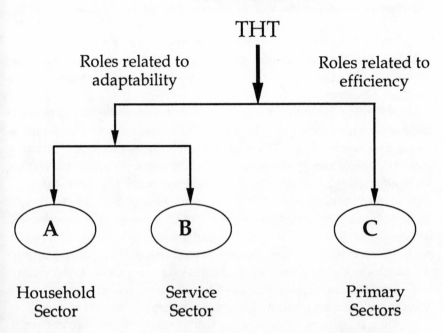

Total Human Time

Figure 6. Socioeconomic structure seen from the control side: Profile of allocation of human time.

    A is the time allocated to non-encoded activities with a net energy consumption (household sector); B is the time allocated to encoded activities with a net energy consumption (service sector); and C is the time allocated to encoded activities with a net energy return (primary economic sectors).

vice sector (B) is less clear because they include both short-term and long-term processes. Nevertheless, we consider them in Figure 6 as FI activities.[8]

Necessary Balance between Efficiency and Adaptability

    Within a defined set of goals and boundary conditions, the existence of different ways of performing the same task is a sign of poor efficiency. In fact, if one of the existing ways is more efficient than the others, keeping alive the others represents an unnecessary cost for the system. To increase the efficiency of the hypercycle, the system should eliminate activities with lower performance as soon as they become obsolete. In other words, given "fixed" goals and boundary conditions, any effort to increase the efficiency of the hypercycle tends to erode diversity.

On the other hand, the system can not "know" if the picture of boundary conditions currently stored in the system of controls is an accurate description of the reality, nor can it know how stable the current boundary conditions are. To anticipate the real possibility of being confronted with different boundary conditions, the system must generate and maintain diversity in its activities. Hence, any effort to increase adaptability tends to decrease efficiency.

On different time scales, it can be seen that adaptability and efficiency depend on each other: On a short time scale (quasi-steady state), only a strong hypercycle (high ET/CI) can generate sufficient surplus to sustain a large fraction of energy throughput for allocation to activities generating variability (a high fraction FI/CI). Clearly, too many activities without a direct return may become a risk by burdening the stability of the energy budget (diminishing the ability to invest in CI to sustain the hypercycle). This is especially true if the system is under stress. In the long term (when dealing with evolution), when boundary conditions will have changed, only the existence of a large diversity of potential states accessible to the society can guarantee that a few of them will be "efficient" under the new conditions to enable an adjustment of the entire system to a new set of activities.

In conclusion, increasing efficiency tends to erode diversity (cost/benefit defined and assessed on a short-term perspective), but it is the only way to sustain diversity in the long term. Adaptability tends to decrease efficiency in the short term, but it is the only way to sustain efficiency in the future.

The Nature of Information Processing

An increased flow of exosomatic energy harnessed by humans, indicated by a higher exosomatic/endosomatic energy ratio, when stabilized in time means that more goods and services are produced and consumed within society, and it means that society as a whole has an increased capability to control and process information. The exo/endo energy ratio of society, defined earlier as the ability to stabilize in time an autocatalytic loop of exosomatic energy, measures the amplification that knowledge and technology induce on the efficacy of human activity. Therefore, it is an emergent property detectable only at the level of society.

Within this frame, human labor provides both applied (muscle) power and information processing. Depending on the type of society, the nature of the contribution of one hour of human labor can vary dramatically between being predominantly power-supply or basically information-processing (Giampietro et al. 1993). The significance of labor in society depends on the cost, accuracy, reliability, and capacity of processing flows of energy and information.

In societies with a low exo/endo energy ratio (poor technology in processing both energy and information), humans perform activities of ecosystem exploitation based on traditional, common knowledge. The flow of information provided by individuals in such societies is highly redundant and, as such, it generates little added value. In pre-industrial societies, the main human contribution in terms of societal control over energy flows is through direct delivery of applied power. Put another way, humans regulate exosomatic energy flows within the autocatalytic exosomatic energy loop by regulating their own flows of muscular power in the autocatalytic endosomatic energy loop. Because of the redundancy in the system of human controls (e.g., farmers performing the same set of activities), the control over a certain activity is roughly proportional to the quantity of human power exerted in that activity. Moreover, the quantity of human energy delivered per unit of time is, in general, quite homogeneous for workers in pre-industrial societies, especially when averaged over the entire year, in spite of differences in gender, age, body size, and nutritional status. This is certainly true when compared to the delivery of machine power (Giampietro et al. 1993). Therefore, in pre-industrial societies, the profile of allocation of endosomatic energy is a good indicator of the allocation of human control over different economic activities. This means that the demand of human control for a particular activity can be assumed to be proportional to the demand of working time for that particular activity.

In the pre-industrial situation, the characteristics of the exosomatic and endosomatic autocatalytic loop are strictly linked. Therefore, physiological and socioeconomic variables are strongly influenced by the nature of the energy budget. A more detailed discussion of the energy budget of pre-industrial societies is provided in Giampietro et al. (1997, this volume).

In societies with a high exo/endo energy ratio (developed countries), workers are mainly providing flexible flows of information for a much more specialized economy powered by machines. Workers, even in the industrial or agricultural sectors, are processing information to direct machines (exosomatic energy flows) that deliver power at levels thousands of times higher than that of human muscle power. The impact of one hour of work in industrialized societies is, therefore, positively related to the exosomatic energy conversions directly performed during work, to the exosomatic energy indirectly required by the amount of capital invested per worker, and to the "quality" of information processing implied in the job. In these societies, the role of humans is no longer related to the delivery of applied power (muscular work) but rather to the processing of information. New technologies, able to better process information (e.g., computers), have further decoupled the amount of labor time from the quantity of information processed in society. Information technology and education have become the main factors in determining the quantity of information that can be processed per hour of labor.

*The Scale Issue: Environmental Loading
and the Need for Adaptability*

The larger the energy throughput (ET) in the system, the larger the size of the process of energy degradation that is based on the destruction of available, favorable biophysical conditions. In practical terms, this means faster depletion of stocks and faster filling of sinks. The resulting stress on boundary conditions is determined by the difference in the speed at which favorable conditions (e.g., gradients of free energy) are generated by biophysical processes and the speed at which favorable conditions are destroyed by the self-organization process of society. After a certain threshold has been reached, the larger the energy throughput of society within a defined ecosystem, the greater will be the stress on boundary conditions and the more pronounced the need for adaptability in the socioeconomic system. Hence, a larger energy throughput requires a higher ratio FI/CI, which is reflected in changes in the pattern of human time allocation (THT/C). This process is confirmed by the current trend of development of human society (see Figures 1-3) and is further illustrated in Giampietro et al. (1997, this volume).

An analysis of the stability of boundary conditions requires a comparison of the scale of the self-organization process of human society, which is proportional to ET, with that of natural ecosystems, which can be assumed to be proportional to the solar energy used in sustaining the self-organization process. Odum (1996) has developed a methodology that can be used to make this comparison. Briefly, the flow of useful work in society (loops of energy that amplify back under the control of humans in the economic process) is compared (after correction for a quality factor) to the flow of useful work in ecosystems (loops of energy that amplify back under the control of biophysical processes). The ratio between the indicator of scale of human activity and the indicator of scale of biophysical activity of ecosystems on which society depends then provides an indicator of *environ-*

mental loading. Odum applies this rationale through a methodology that estimates indicators in terms of EMergy assessments[9] This operationalization is still controversial.

Nevertheless, approximations can be made of the requirement of an economy for ecological services in terms of a requirement of space-time of the biophysical activity needed to make available the inputs of matter and energy to the economic process and to absorb the related outputs. Possible approaches, following Odum's methodology, include: defining the amount of inputs and/or wastes, assessing the space-time requirement per unit of input and waste, and checking whether sufficient biophysical activity is actually available within the area occupied by society to sustain its economy without degrading the environment. Examples are the "BIOSTA budget" (Giampietro and Pimentel 1991b) and the "ecological footprint" (Rees and Wackernagel 1994).

When society's requirements exceed available ecological services, the gap can be filled through importation by using the embodied space-time activity of other ecosystems that are distant in time (e.g., oil stocks) or in space (e.g., imported commodities or animal feed). Another, temporary solution is overdrafting the ecological services available, which will steadily erode the resilience of local ecosystems and eventually destroy the complexity of natural systems.

In conclusion, using the rationale proposed by Odum, the intensity of the activity of self-organization of human society, defined as the ratio ET/ exploited area, can be used as an indicator of environmental loading. Such an indicator can then be related to the increase of negative effects caused by human activity on the environment. Based on cases where this has happened, a critical environmental loading (ELmax) can be defined that represents the threshold value for ecological compatibility. When the intensity of human activity exceeds that value, negative consequences are likely to appear.

PUTTING THE PIECES TOGETHER

I will now put together the building blocks of the model by reorganizing the basic ideas and concepts in a series of points and schemes.

The Basic Autocatalytic Loop of Endosomatic Energy

The ability of society to stabilize its process of energy dissipation in time is related to the generation of an autocatalytic loop of useful energy. Useful energy is invested by society in making available the inputs of matter and energy consumed by the system.

The simplest autocatalytic loop of useful energy is endosomatic. This is the principal loop observed (in steady state analyses) in pre-industrial society. In pre-industrial society, human beings are able to obtain a reliable supply of food, of energy useful for cooking, shelter, and clothing, and of other ecological services, as a result of their (mainly muscular) work and the work of their surrounding ecosystems. The efficacy of the endosomatic loop is amplified by the use of exosomatic energy, such as fire and animal, wind, and water power. However, in these societies, the control over exosomatic energy flows is too primitive to generate an autocatalytic loop in terms of exosomatic energy, so the overall exo/endo energy ratio seldom exceeds 5/1 (Giampietro et al. 1993). Under these circumstances and within a defined set of boundary conditions (defined by the quality and quantity of exploited ecosystem and available technology), the pattern of human time allocation defines the pattern of allocation of available power supply and is the key factor in determining the feasibility and characteristics of the autocatalytic loop of endosomatic energy.

The simple endosomatic autocatalytic loop can alternatively be described from a hierarchical perspective in terms of division of human control over efficiency and adaptability, making use of the distinction between A, B, and C activities. A "triadic" reading (see Salthe 1985) is illustrated in Figure 7, which describes the allocation of human control (time/activity) available to the socioeconomic system on three levels: the focus, upper and lower level. This example shows that humans have to pay a tribute, in the form of time allocation, to all three hierarchical levels.

A tribute is paid to the higher hierarchical level in the form of A activities. This tribute is necessary to guarantee adaptability in the long term. This higher level relates to society in a historic perspective: A activities provide society with initiating conditions, including the reproduction of humans, cultural identity, knowledge, and technological capital. A tribute is paid to the focus level in the form of B activities to ensure the everyday maintenance of the structure of human mass. B activities provide the system of controls over the network of matter and energy consumed by society. A tribute is also paid to the lower hierarchical level in the form of C activities to guarantee efficiency in the set of everyday operations. C activities guarantee the needed flow of inputs from the environment.

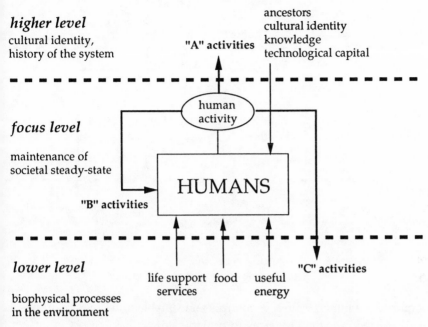

higher level
cultural identity,
history of the system
　　　　　　　　　　　"A" activities
ancestors
cultural identity
knowledge
technological capital

focus level

maintenance of
societal steady-state
　　　　　　　　　　human activity
　　　　　　　　　　HUMANS
"B" activities

lower level
biophysical processes
in the environment
　　　　life support　food　useful　"C" activities
　　　　services　　　　　　energy

Figure 7.　Endosomatic autocatalytic loop: A hierarchical view
of the allocation of controls by society.

Emergence of an Autocatalytic Loop of Exosomatic Energy

The endosomatic autocatalytic loop of subsistence societies can be described as: food → human work → more food → more human work. Intrinsic physiological limitations related to human metabolism constrain the expansion of this positive feed back loop (see Giampietro et al. 1997, this volume). At the level of society, the intrinsic limitations to the speed of dissipation of endosomatic energy induce limits to the speed of dissipation of exosomatic energy. These constraints were only overcome with the Industrial Revolution, when human societies introduced a new positive feedback loop: machines → access to fossil energy → more machines → access to more fossil energy.

The autocatalytic loop of exosomatic energy is observed in industrial societies. Humans in industrial societies are able to obtain a reliable flow of food and useful energy for the production and consumption of services, goods, and ecological services thanks to the useful energy generated by exosomatic devices and the work of natural ecosystems. Note that an auto-catalytic loop of exosomatic energy does not operate independently but

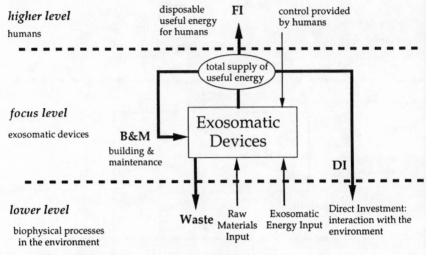

Figure 8.   Exosomatic autocatalytic loop: A hierarchical view
of the allocation of useful energy by society.

requires that humans provide controls and suitable initiating conditions as
well as environmental services to provide suitable sources of inputs and
sinks.

When describing the autocatalytic exosomatic loop from a hierarchical
perspective, as illustrated in Figure 8, we find that exosomatic devices, too,
have to pay a tribute to the hierarchical levels involved in their existence
and operation. This tribute is paid in the form of FI useful energy. This is
the disposable useful energy that humans obtain in return for the construction and maintenance of machines. Further, a tribute is paid to the focus
level in the form of B&M. This is the useful energy generated by machines
needed to build and maintain their own structure. A third tribute of useful
energy is paid to the lower hierarchical level. This is the useful energy generated by machines needed for the stabilization of inputs of matter and
energy and the disposal of waste into the environment. It includes the
energy spent in activities needed to guarantee energy and mining, and food
and environmental security.

The autocatalytic loop generated by machines has two distinct interfaces
(see Figure 9), one with humans and one with the environment. Regarding
the former, machines represent a cost for humans in terms of human labor
demand but a return in terms of a net supply of useful energy for human
use. Regarding the latter, machines alter boundary conditions with their
activity through the withdrawal of inputs and disposal of wastes. As noted

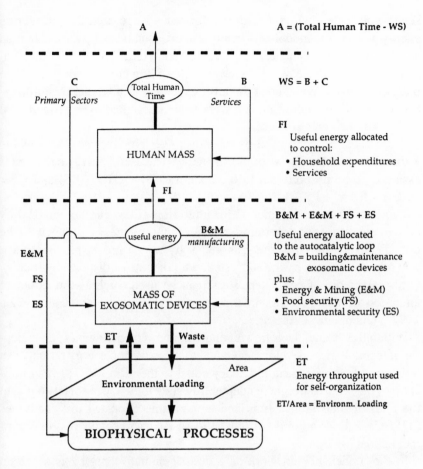

Figure 9. Biophysical constraints on the resonance between "controls generating useful energy" and "useful energy generating controls."

earlier, given a particular geographical area, the scale of machine activity (energy conversions controlled by machines) relative to the scale of ecological activity (energy conversions controlled by ecosystems) defines a certain environmental loading ratio.

The Double Autocatalytic Loop of Modern Society

Modern societies are characterized by the existence of a double autocatalytic loop, as illustrated in Figure 9: an autocatalytic loop of endosomatic energy combined with an autocatalytic loop of exosomatic energy. The scheme presented in Figure 9 further illustrates the reasons for considering energy flows in the model only in terms of exosomatic energy. First, human control over exosomatic devices is estimated in terms of human time allocation to work. Hence, work is viewed as providing information rather than muscular power. Averaged at the level of society as a whole, the information delivered per unit of human work time is assumed to be proportional to the exo/endo energy ratio.

Second, controls provided by machines over biophysical processes are estimated in terms of flows of useful energy. The useful energy delivered is assumed to be proportional to the quantities FI = (HH + SS) and CI = (B&M + E&M + FS + ES).

Third and last, environmental loading ratios (ELs) can be obtained by comparing the rate of generation of major inputs by the ecosystem with the rate of consumption by society, as well as by comparing the rate of absorption of major wastes by the ecosystem with the rate of disposal by society. Other indicators of ecosystem stress can be used, such as the extent of alteration of energy flows in the ecosystem (Giampietro et al. 1992a, 1992b; Giampietro in press).

Within this frame, the following indicators can be defined: (1) an indicator of technological development equal to the exo/endo energy ratio; (2) an indicator of technological efficiency equal to the ratio FI/ET; (3) an indicator of the relative weight of adaptability and efficiency in society equal to the ratio of total human time and human time allocated to C activities (THT/C); and (4) a family of indicators of environmental stress (environmental loading indices).

In the model, human work as a unit of human control has a certain opportunity cost in terms of demand of useful energy from the exosomatic hypercyclic compartment. This opportunity cost is proportional to the material standard of living in society as measured by the demand of FI per unit of work supply. In the same way, a unit of exosomatic useful energy supplied by the hypercyclic compartment has an opportunity cost in terms of demand of work from the human compartment. This opportunity cost is proportional to the strength of the hypercycle generating the flow of useful energy. Figure 10 illustrates how the double autocatalytic loop resonates between human activity and useful energy: The endosomatic loop generates useful energy in the exosomatic loop on one time scale, and useful energy in the exosomatic loop generates human activity in the endosomatic loop on a different time scale.

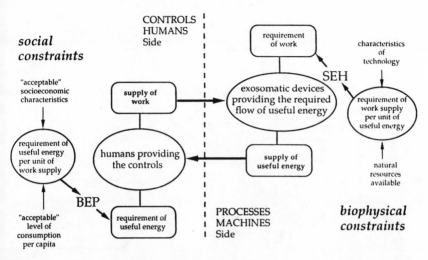

Figure 10. Dynamic interaction between allocation of controls and useful energy in society.

BEP is the bioeconomic pressure or energy consumption per hour of labor in the primary economic sectors. SEH is the strength of the exosomatic hypercycle or supply of useful energy per hour of labor in the primary economic sectors.

Dynamic Equilibrium between Endosomatic and Exosomatic Loops

The use of exosomatic energy (processes) and human time (controls) in modern society is schematized in Figure 11. (The case of pre-industrial societies is simpler and is presented in Giampietro et al. 1997, this volume). The exosomatic useful energy is used in CI activities to procure and transform energy input in the energy sector and to procure raw materials (E&M), to build and maintain exosomatic devices in the manufacturing sector (B&M), to guarantee food security in the food system (FS), and to guarantee environmental security (ES), as well as in FI activities to provide services in the service sector (SS) and to support human activity outside work in the household sector (HH). The first four CI activities together require an amount of human time equal to C, whereas the latter two FI activities absorb an amount of human time equal to B and A, respectively. Note that the total human time available in society (THT) equals the sum of the time allocated to A, B, and C activities, whereas the time available for work (work supply or WS) equals the sum of the time allocated to B and C activities.

In conclusion, by assessing the flows and parameters presented in Figure 11, we can concisely describe, for any defined society, the autocatalytic loop of exosomatic energy in terms of the overall requirement of human

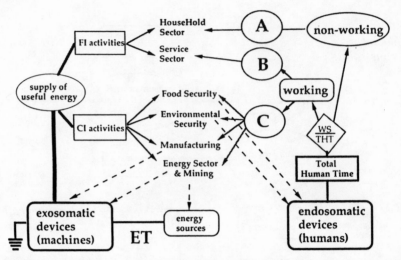

Figure 11. Parallel allocation of exosomatic energy and human time in society.

work to run the primary sectors of the economy, and the overall exosomatic energy consumption of the society as a whole.

This analysis establishes a bridge between a socioeconomic reading of the development of society (based on changes in socioeconomic variables) and a biophysical reading of the same process (based on changes in the technical coefficients describing the autocatalytic loop of exosomatic energy). In fact, the balancing of the energy budget of a society implies that the exosomatic energy consumption of society as a whole per hour of work in the hypercycle (this parameter is called bioeconomic pressure or BEP) must be matched by the exosomatic energy supplied by the hypercycle per hour of work required there (this parameter is called strength of the exosomatic hypercycle, or SEH) (see Figure 10). The demand side of the energy budget—that is, the energy throughput ET consumed by society per hour of labor delivered in the primary sectors of the economy (BEP)—is defined by variables that refer to a socioeconomic description of society (see Figure 12). On the other hand, the supply side of the energy budget—that is, the energy throughput ET delivered to society per unit of work time required in the primary sectors of the economy (SEH)—is defined by the technical coefficients of the various sectors involved (building and maintenance, energy and mining, food security, and environmental security; see Figure 13).

Thus, the two readings of the socioeconomic system illustrated in Figures 12 and 13 establish a link among different classes of variables: phys-

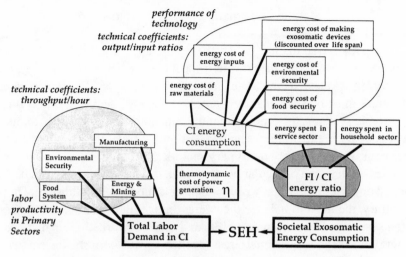

Figure 12.   Variables defining the bioeconomic pressure BEP.

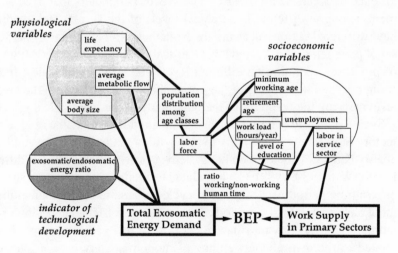

Figure 13.   Variables defining the strength of the hypercycle SEH.

iological variables, such as body size and metabolic flow (Figure 12); socioeconomic variables, such as size of labor force and work load (Figure 12); technical coefficients (Figures 11 and 13); and indicators of environmental stress (Figure 9). Equations that link these variables are provided in the next section, with numerical examples provided in Giampietro et al. (1997, this volume).

Stability of Dynamic Equilibrium in Time

Dissipative systems are not in thermodynamic equilibrium. However, this does not mean that they do not respect the laws of thermodynamics. To remain in a quasi-steady state, the metabolism typical of the system's structure must be sustained by a process of exergy degradation. In socioeconomic systems, this implies that the consumption of energy per hour of work supply in the primary sectors (BEP) must equal the supply of energy per hour of work supply in primary sectors (SEH).

The fact that, at a particular point in time, the energy budget must be balanced does not imply that we are adopting a model of equilibrium. On the contrary, the need to balance the energy budget in conditions of systemic stress (such as Malthusian instability and cyclic perturbations in matter and energy flows) and occasional stress (such as acyclic perturbations in matter and energy flows), requires that the system continuously adjust the various sets of parameters, on the control and the process sides, to maintain congruence between BEP and SEH. The sets of parameters that need adjustment belong to different hierarchical levels of the system and, therefore, have different lag times in changing. Nonetheless, changes in any particular set of parameters at a defined level and scale affect all the other parameters of processes operating on different levels and space-time scales. Indeed, sooner or later, the effects of the initial changes will return to the original level carrying the feedback of the other parts of the hierarchical system.

Such a dynamic energy budget will be anything but stable, and the trajectory of its evolution is likely to be anything but linear. For example, the energy budget of socioeconomic systems is affected by cycles of different length (Watt 1989, 1992): Cycles related to changes in population structure have a time period of 20-25 years; cycles related to technological investment have a shorter time period of about 5-7 years. Changes in laws regulating modern society may have a lag time of few years, whereas changes related to the institutional settings of modern society have a longer lag time, such as changes in cultural identity. Finally, there are cycles related to changes in metastable equilibria of ecological processes that, undoubtedly, have longer time periods. Therefore, dynamic balancing of the energy budget, as implied by this model, means that in the short term, the configuration of the socioeconomic system tends to remain around attractor solutions. It will move through various combinations of values for variables that determine congruence between BEP and SEH and that are allowed by the existing system of controls.

The result will be a pulsing behavior that does not reach any fixed equilibrium point (see Giampietro et al. 1997, this volume). When adjustments in the slowest set of parameters finally move the system closer to the original equilibrium points, a number of irreversible events (changes in internal characteristics of the system and boundary conditions) will have accumulated in the system to a point that the original basin of attraction is no longer viable. Indeed, the internal evolution of structures and functions of society (such as an increase in the minimum acceptable material standard of living) and/or changes in the available gradients of free energy (such as a decrease in resource quality when most favorable gradients are exploited first) in time will change the set of possible combinations of parameters that provide a dynamic equilibrium between BEP and SEH.

On a long time scale, rather than smooth trajectories of evolution, we should expect to see periods of stability in which the system remains in its current basin of attraction, oscillating around average values of population density (which determine correspondent values of Environmental Loadings) and material standard of living (BEP), followed by abrupt changes when the system falls into a different basin of attraction. These abrupt changes—the so-called "revolutions" in human civilization, such as the agricultural revolution, the industrial revolution, and the computer revolution—are due to a simultaneous rearrangement of parameters that describe the characteristics of the system on different levels. By this is meant a simultaneous change in population and/or material standard of living (BEP) on the one hand, and changes in the nature of the hypercyclic compartment (inducing dramatic changes in SEH) on the other hand. With a simultaneous rearrangement of parameters around new values of the equilibrium between BEP and SEH and a new value of EL, the system "jumps" into a new attractor solution with a new set of accessible steady states for its socioeconomic organization.

THE EQUATIONS OF THE MODEL

Energy Budget from the Demand Side

Bioeconomic Pressure

Bioeconomic pressure (BEP) is defined as the exosomatic energy throughput consumed at the level of society per unit of labor time in the primary economic sectors. In Figure 12, several variables that characterize the socioeconomic organization of society are indicated as determining the bioeconomic pressure (BEP), including per capita energy consumption, the ratio non-labor to labor time, and the percentage of the labor force engaged in the service sector. The combination of the values that these variables take on at a defined point in time defines a value for the bioeconomic pressure (BEP) in society.

At equilibrium between BEP and SEH, the bioeconomic pressure can be seen as the *opportunity cost of human labor* in a given socioeconomic system and, as such, establishes a link between the standard of living, technology, and natural resources. Indeed, the term "bioeconomic pressure" was suggested by Franck-Dominique Vivien to refer to Georgescu-Roegen's idea that an increase in intensity of the economic process induces, as a biophysical side-effect, an increase in the intensity of the throughputs of energy in the primary economic (CI) sectors.

Bioeconomic Pressure as an Indicator of Standard of Living

We now come to the equations that show how the bioeconomic pressure reflects the material standard of living in society.

The requirement of exosomatic energy of society can be written as:

$$ET = (MF \times ABM) \times Exo/Endo \times THT \tag{1}$$

The supply of exosomatic energy of society can be written as:

$$ET = WS \times [C/(B+C)] \times BEP = C \times BEP \tag{2}$$

Where:

ET = the energy throughput defined as the flow of exosomatic energy in society (J/yr),

MF = the metabolic flow defined as the flow of metabolic energy per kg of body mass ($J\ kg^{-1}\ hr^{-1}$),

ABM = the average body mass defined as the total mass (kg) of population divided by the number of individuals in society (data available from FAO statistics),

$Exo/Endo$ = the ratio between exosomatic and endosomatic energy flows in society,

THT = the total human time defined as the number of individuals (population size) multiplied by the number of hours in a year (*hr*),

WS = the work supply defined as the amount of time (*hr*) that the economically active population allocates to work on a yearly basis (as opposed to sleeping, leisure, etc.),

C = the number of hours of work delivered in the CI sectors of the economy on a yearly basis (*hr*),

B = the number of hours of work delivered in the service sector of the economy on a yearly basis (*hr*), and

WS = *B + C*.

By combining Equations (1) and (2), we obtain:

$$BEP = ET/C = (ABM \times MF) \times Exo/Endo \times THT/C \tag{3}$$

Numerical examples of ranges of values for MF, ABM, Exo/Endo, THT/C, BEP are given in Table 1. The possible use of BEP as an indicator of material standard of living is discussed in Giampietro et al. (1997, this volume).

Energy Budget from the Supply Side

Strength of the Exosomatic Hypercycle (SEH)

The exosomatic autocatalytic loop does not per se determine the amount of energy consumed by society. However, its internal characteristics do define a relation between standard of living (BEP) and ecological stress per unit of population mass (Environmental Loading [EL]; see Figure 9). This means that technological improvement in the efficiency of exosomatic devices can be used by society (at a fixed level of population) to augment the standard of living (BEP) while keeping the same environmental loading (EL), or to reduce the environmental loading without having to reduce BEP (see Giampietro et al. [1997] for a discussion on this dilemma). The factors describing the relation between exosomatic energy consumption and human work are illustrated in Figure 13.

The factors determining consumption of exosomatic energy in the CI compartment can be described as follows (see Figure 11):

$$CI = \Sigma \, CI_i = B\&M + E\&M + ES + FS \tag{4}$$

Table 1. Metabolic Flow (MF), Average Body Mass (ABM), "ABM × MF," Exo/Endo Energy Ratio, THT/C, and BEP for Selected Countries

Country	MF kJ/kg/hour	ABM kg	ABM × MF MJ/hour/capita	Exo/Endo Energy Ratio	THT/C	BEP MJ/hour
Canada	7.11	55.3	0.39	89.6	41.9	1476.7
United States	6.97	57.7	0.40	89.3	37.3	1339.8
Belgium	6.85	59.4	0.41	56.2	41.0	936.9
Australia	7.26	54.0	0.39	63.7	36.9	921.5
The Netherlands	6.89	59.3	0.41	58.2	38.1	906.5
Norway	6.94	57.3	0.40	56.4	34.8	780.4
France	7.08	55.4	0.39	47.9	36.4	684.2
United Kingdom	6.95	56.8	0.39	45.8	37.1	670.6
Italy	6.91	58.2	0.40	33.9	31.8	432.7
Switzerland	6.90	58.0	0.40	39.4	26.7	420.7
Spain	7.29	53.3	0.39	26.5	36.1	372.1
Argentina	7.55	50.7	0.38	17.5	34.3	230.0
Mexico	8.39	42.1	0.35	17.8	25.5	160.6
Egypt	8.48	44.7	0.38	5.8	28.1	61.3
Bolivia	8.82	39.7	0.35	5.0	26.2	45.9
Philippines	9.14	37.7	0.34	5.0	20.2	34.2
India	9.25	36.6	0.34	5.0	18.5	31.4
Bangladesh	9.51	34.3	0.33	5.0	16.9	27.6
Nigeria	9.02	39.2	0.35	5.0	15.5	27.4
China	8.48	43.0	0.37	7.6	9.3	26.0
Senegal	8.80	40.6	0.36	5.0	14.1	25.2
Ethiopia	9.31	35.5	0.33	5.0	15.0	24.8
Burkina Faso	8.83	40.8	0.36	5.0	12.6	22.7
Niger	8.89	39.7	0.35	5.0	12.3	21.7
Burundi	9.25	36.6	0.34	5.0	10.6	18.0

Source: Data from Pastore et al. (1996).

120

To characterize the technological performance in each of the four primary sectors, I introduce the following family of indicators of efficiency, or "quality," of technological processes:

Energetic cost of energy and mineral resources. This may be defined as the fraction of the energy throughput in society spent in energy and mining (E&M/ET). Alternatively, it may be defined in terms of input/output ratios—that is, the energy spent per unit of energy input available to society (e.g., the EROI of energy sources) and the energy spent per kg of minerals made available to the economy (e.g., Hall et al. 1986).

Energetic cost of building exosomatic devices. This may be defined as the fraction of energy throughput in society spent in building and maintenance (B&M/ET). Alternatively, it may be defined in terms of an input/throughput ratio—that is, the energy spent in building and maintaining the exosomatic devices, discounted over their life spans, per unit of energy consumed in their operation.

Energetic cost of food security. This may be defined as the fraction of the energy throughput in society spent in the food system (FS/ET). Alternatively, it may be defined as an exo/endo energy ratio—that is, the exosomatic energy spent in the food system per unit of food energy consumed by humans in society.

Energetic cost of environmental security. This may be defined as the fraction of the energy throughput in society that is spent in environmental security (ES/ET). Alternatively, it may be defined as an exo/endo energy ratio—that is, the exosomatic energy spent in environmental security per unit of endosomatic energy consumed by the population.

Energy for transportation (generally reported as a separate item in energy statistics) is redistributed in this model to the various sectors (for details, see Pastore et al. 1996).

The productivity of labor (CI/C) in the four primary sectors CI_i may be defined as the energy throughput (J) spent in sector CI_i per hour of work delivered there (ET_{CI_i}/h). The number of hours of work in the entire CI sector (C) can be expressed as follows

$$C = \frac{B\&M}{ET_{B\&M/h}} + \frac{E\&M}{ET_{E\&M/h}} + \frac{FS}{ET_{FS/h}} + \frac{FS}{ET_{FS/h}} + \frac{ES}{ET_{ES/h}} \qquad (5)$$

The Strength of the Exosomatic Hypercycle as Defined by Technical Coefficients

According to Equations (4) and (5), we can write:

$$CI/C = \frac{B\&M + E\&M + FS + ES}{\dfrac{B\&M}{ET_{B\&M/h}} + \dfrac{E\&M}{ET_{E\&M/h}} + \dfrac{FS}{ET_{FS/h}} + \dfrac{ES}{ET_{ES/h}}} \tag{6}$$

that leads to:

$$SEH = \frac{B\&M + E\&M + FS + ES}{\dfrac{B\&M}{ET_{B\&M/h}} + \dfrac{E\&M}{ET_{E\&M/h}} + \dfrac{FS}{ET_{FS/h}} + \dfrac{ES}{ET_{ES/h}}} \times \frac{ET}{CI} \tag{7}$$

that can be written in a simpler form as:

$$SEH = ET/C = ET/CI \times CI/C \tag{8}$$

Equation (8) illustrates clearly that the strength of the hypercycle depends on two types of energy efficiency for technology: a higher output/input ratio ("type 1" efficiency), leading to an increase in the ratio ET/CI, and a higher throughput/hour ("type 2" efficiency), leading to an increase in the ratio CI/C (for more, see Mayumi 1991; Giampietro et al. 1997, this volume). An increase in either efficiency implies a decrease in the other, according to the maximum power principle (Odum and Pinkerton 1955).

Internal Biophysical Constraints on the Balancing of the Energy Budget

Equations (3) and (8) show that to have a feasible solution for the energy budget of a particular society seen at a particular moment in time (a quasi-steady state configuration), the parameters determining BEP must maintain congruence with changes in the variables determining SEH. Thus:

$$BEP = SEH \tag{9}$$

This link has two important implications. First, the ability to achieve a certain value of SEH is determined by technological and/or ecological parameters. In other words, the stability of a steady state operating at a certain level of SEH depends on biophysical constraints related to the strength of the hypercycle of exosomatic energy. Indeed, at a low exo/endo ratio, when the autocatalytic loop of exosomatic energy is weak, the only way to stabilize the budget is by adjusting socioeconomic variables. In this case, parameters determining SEH (technical coefficients defining the characteristics of the exosomatic loop) force the physiological and socioeconomic variables to take on values that give a low BEP (e.g., in the range of 10 MJ/hour), even if this implies a poor material standard of living.

For instance, when human power is the main system of generating useful energy, an increase in the fraction of elderly in the population reduces the number of labor hours and the power supply (the effectiveness per hour of labor) in society. Indeed, without heavy reliance on exosomatic energy flows, improvements in the standard of living can destabilize the energy budget (Giampietro et al. 1993). Thus, the required equality of BEP and SEH means that the human time allocated to the economic process must produce enough energy surplus to sustain not only the consumption of the labor mass outside work hours but also the consumption of the economically inactive population mass composed of elderly, children, unemployed, students, and disabled. In developed countries, such as the United States, each hour of labor must sustain nine hours of human time that are entirely devoted to consumption (A activities) (Giampietro et al. 1993). Hence, societies with a weak hypercycle (SEH) can not afford a large fraction of nonproductive population mass. Often, the only "idle" population mass present are children needed for replacement (Giampietro et al. 1993). Examine, for example, the Yanomamö society in Figure 1.

Second, the willingness of individuals to remain in a society operating at a defined value of BEP depends on their cultural identity and system of values. Therefore, the stability of a particular energy budget (the configuration of parameters determining equality of BEP and SEH) depends also on cultural identity associated with the socioeconomic system. In general, we can expect a certain "pressure" (BEP) within society exerted by individual human beings—and, thus, generated at the lower hierarchical level—for a continuous improvement in the standard of living. This is especially true for young people and parents of young children in developing countries. The strength of this pressure is directly related to what people conceive as a minimum "acceptable" standard of living. Hence, access to information about different social systems that have a better material standard of living can dramatically affect what is considered an "acceptable minimum BEP" in society. In the last decades, access to information about the levels of BEP achieved in highly industrialized countries has generated instability in poor and traditional societies operating at a much lower BEP. Individual members of less developed societies tend to abandon the traditional pattern of exploitation of natural resources (the traditional set of codified roles based on accumulated experience in the system of social controls) in an attempt to switch to a different set of activities that promise to provide a higher return for human labor (higher SEH). Driven by individual aspirations and fueled by gradients of wealth within society and among societies, the continuous struggle for an increase in BEP necessarily implies a struggle to continuously increase SEH through improved technological performance. This intrinsic mechanism pushing socioeconomic systems toward areas of instability is discussed in the next chapter (Giampietro et al. 1997) in relation to the general drive of complex dissipative adaptive systems to become something else.

CONCLUSIONS

Socioeconomic systems are complex systems that can be characterized as dissipative, holarchic, and adaptive. They are dissipative because societies are self-organizing systems based on a turnover process of their structure, which is sustained by a process of exergy degradation. They are holarchic because they are made up of holons, individual entities with a structure and a functional role, that can be defined only on different space-time scales. They are adaptive because in the long term, socioeconomic systems tend to become something else: Changes in the internal system of entailments or changes in boundary conditions will, sooner or later, make the old pattern of interaction with the environment no longer feasible. Maximizing efficiency (a short-term perspective) and maximizing adaptability (a long-term perspective) in society imply contrasting strategies, even though efficiency and adaptability depend on each other, on different time scales.

Given these characteristics, socioeconomic systems can not be described by adopting a single perspective. Certainly, reductionist analyses of the interaction between society and nature can be useful in dealing with specific problems where the ceteris paribus hypothesis can be applied. However, forcing the description of complex systems operating on multiple scales into one single reading inevitably results in the loss of essential information about the system. Complexity implies the use of multiple spatiotemporal scales and contrasting perspectives. Consequently, uncertainty and value commitments are unavoidable characteristics of any analysis of socioeconomic systems, in particular when the issue of sustainability is addressed. Evolutionary paths of the system have to be evaluated and related to situations defined at particular points in time and space. In this process, scientists face the challenge of dealing with self-fulfilling prophecies, nonlinear changes in boundary conditions, contrasting perspectives obtained from different holons belonging to the same system, and unknown external perturbations in the future. Rather than pretending that these difficulties do not exist, scientists should make an effort to make themselves useful. They should try to bridge their own reading of socioeconomic systems (after acknowledging the fact that their choice of a particular perspective implies losing a lot of valuable information) to the reading given by other scientists.

Models of biophysical analyses of socioeconomic systems can be used to establish bridges among different perspectives on the process of technological development of society. Different sets of numerical indicators can be used to assess contrasting effects of the same change when perceived on different hierarchical levels. Clearly, these models can *not* be used to define what is "better" or "worse" for the system, predict what the system will do, or calculate a unique set of costs and benefits generated by a particular change to enable a "passion-free" rational process of decision making. The correct uses proposed for biophysical models of analysis are threefold. One is to link different descriptions of socioeconomic systems performed at different hierarchical levels (e.g., linking the perspectives of demographers, nutritonists, economists, sociologists, anthropologists, philosophers, etc.). A second is to analyze and discuss trade-offs implied by changes in socioeconomic systems in terms of efficiency and adaptability. A third is to study the nature of, and the mechanisms determining, the process of reciprocal entailment of different systems of entailments within human societies (how cultural identity affects economic processes and how economic processes affect culture).

Clearly, the interpretation given to the indicators obtained by such a biophysical analysis will still depend on the cultural identity, position in the social hierarchy of power, and system of values of the person performing the analysis. Any interpretation of indicators obtained from a given set of parameters versus indicators obtained from a different set (concerning ecological integrity versus equity, material standard of living, and what is perceived to be quality of life) is and remains, in spite of the use of equations, value-dependent. The general framework of analysis, based on different sets of biophysical indicators capable of bridging different perspectives of development, should avoid dangerous misunderstandings when scientists of different disciplines talk about the same system viewed from different windows of observation.

Finally, another use of this model is proposed in the next paper (Giampietro et al. 1997) as a tool for "tutoring" the fashionable activity of "envisioning" our common future. If it is true that human society is heavily affected by "passional entailment" (that humans generate self-fulfilling prophecies), it is also true that not everything that we want to do can actually be done. Thermodynamic constraints do affect the feasibility of future paths of development. Thus, the model can be used to check the feasibility of the energy budget of "envisioned societies."

NOTES

1. The unit of energy used in this paper is the joule (J). 1 kJ (kilojoule) = 10^3 J; 1 MJ (megajoule) = 10^6 J; 1 GJ (gigajoule) = 10^9 J; 1 TJ (terajoule) = 10^{12} J. For conversion purposes, 1 kcal = 4.186 kJ.

2. Exergy is a measure of the free energy available to the system. Exergy can be considered the fuel for any system that converts energy and matter in a metabolic process. For more on exergy, see Ahern (1980), Pillet et al. (1987), and Jørgensen (1992).

3. Clearly, this is an approximation since measures of "quality" of energy input are involved. For instance, a unit of energy input can provide different quantities of useful energy depending on how it is used. Hence, the ratio "energy input consumed/useful energy generated" can change in time according to the efficiency of available technologies (e.g., the efficiency of using coal to generate electric power) and the mix of energy sources used (e.g., the relative contribution of coal, oil and natural gas in producing electricity). Nevertheless, at any particular moment in time (at a certain level of technology), we can calculate an energy input equivalent (by introducing a quality factor for different types of energy input) that can be assumed proportional to the useful energy dissipated. For example, in U.N. statistics, all consumption is given in oil equivalents. For instance, 1 MJ of electricity obtained from a hydroelectric plant is considered to be equivalent to 3 MJ of oil, that is, the quantity of oil required to generate 1 MJ of electricity with current technology.

4. Even though the sector of manufacturing is a net consumer of energy in standard input/output analyses of economic sectors, in our analysis the building of exosomatic devices (e.g., machines able to convert energy input into useful energy outside the human body) is an activity essential to the establishment of the hypercycle based on machines and fossil energy (autocatalytic loop of exosomatic energy).

5. The terms "endosomatic" and "exosomatic" were proposed by Georgescu-Roegen (1975) to describe the idea, originally stated by Lotka, that energy spent in society under human control, even if converted outside the human body, should be considered as an expanded form of human metabolism: "it has in a most real way bound men together into one body: so very real and material is the bond that society might aptly be described as one huge multiple Siamese twin" (Lotka 1956, p. 369). The vivid image proposed by Lotka explicitly suggests that a hierarchical level of organization higher than the individual should be considered to describe the flow of energy in modern societies.

6. The citation of Georgescu-Roegen reads: "If the count shows that a box contains twenty candies, we can make twenty youngsters happy now or tomorrow, or some today

and others tomorrow, and so on. But...the case of an electric bulb which lasts five hundred hours. We can not use it to light five hundred rooms for an hour now." *Stocks* (e.g., candies) are not limited in their flow dimension; however, the flow dimension at which we decide to exploit a stock will affect the lifespan of the stock in time. To the contrary, *funds* (e.g., light bulbs) are severely limited in their flow dimension; however, when their natural flow dimension is respected, they can be assumed to be unlimited in their temporal dimension (at least if 500 hours are more than the time scale considered; the sun would be a better example than the light bulb).

7. By "restriation," Zipf means a different pattern of allocation in space and time of matter and energy flows.

8. For the sake of simplicity, I adopt the definition of A and B activities as FI activities, and C activities as CI activities (as in Figure 5). However, in terms of roles, activities performed in the service sector are "in between" FI and CI investments.

9. Natural transformations within ecological systems can be assumed constant when assessed on the spatiotemporal scale at which economic processes are operating. For instance, the solar energy needed to make 1 kg of wood or to make available 1 kg of rain can be assumed constant and defined according to the organizational pattern of an ecosystem when analyzed on an economic time scale. Therefore, a unit of a particular natural resource needed by society from a particular ecosystem can be related to a defined demand of space and time of its biophysical activity.

REFERENCES

Adams, R.N. 1988. *The Eighth Day: Social Evolution as the Self-Organization of Energy.* Austin: University of Texas Press.

Ahern, J.I. 1980. *The Exergy Method of Energy Systems Analysis.* New York: John Wiley & Sons.

Allen, T.F.H., and T.W. Hoekstra 1992. *Toward a Unified Ecology.* New York: Columbia University Press.

Allen, T.F.H., and T.B. Starr 1982. *Hierarchy.* Chicago: The University of Chicago Press.

Bailey, K.D. 1990. *Social Entropy Theory.* Albany: State University of New York Press.

Barham, J. 1990. "A Poincaréan approach to evolutionary epistemology." *Journal of Social and Biological Structures* 13: 193-258.

Brooks, D.R., J. Collier, B.A. Maurer, J.D.H. Smith, and E.O. Wiley. 1989. "Entropy and Information in Evolving Biological Systems." *Biology and Philosophy* 4: 407-432.

Brooks, D.R., and E.O. Wiley 1988. *Evolution as Entropy.* Chicago: The University of Chicago Press.

Carlstein, T. 1982. "Time Resources, Society and Ecology: On the Capacity for Human Interaction in Space and Time in Preindustrial Societies." *Lund Studies in Geography*, Ser. B: *Human Geography* No. 49. Lund: The Royal University of Lund.

Conrad, M. 1983. *Adaptability: The Significance of Variability from Molecule to Ecosystem.* New York: Plenum.

Cottrell, W.F. 1955. *Energy and Society: The Relation between Energy, Social Change and Economic Development.* New York: McGraw-Hill.

Debeir, J-C., J-P. Deléage, and D. Hémery 1991. *In the Servitude of Power: Energy and Civilization through the Ages.* Atlantic Highlands, NJ: Zed Books.

Depew, D.J., and B.H. Weber, eds. 1985. *Evolution at a Crossroads: The New Biology and the New Philosophy of Science*. Cambridge, MA: MIT Press.

Depew, D.J., and B.H. Weber. 1994. *Darwinism Evolving: Systems Dynamics and the Genealogy of Natural Selection*. Cambridge, MA: MIT Press.

Funtowicz, S.O., and J.R. Ravetz 1990. *Uncertainty and Quality in Science for Policy*. Dordrecht: Kluwer Academic Publishers.

———. 1994. "The Worth of a Songbird: Ecological Economics as a Post-normal Science." *Ecological Economics* 10(3): 197-207.

Georgescu-Roegen, N. 1971. *The Entropy Law and the Economic Process*. Cambridge, MA: Harvard University Press.

———. 1975. "Energy and Economic Myths." *Southern Economic Journal* 41: 347-381.

Gever, J., R. Kaufmann, D. Skole, and C. Vörösmarty 1991. *Beyond Oil: The Threat to Food and Fuel in the Coming Decades*. Niwot: University Press of Colorado.

Giampietro, M. 1994a. "Sustainability and Technological Development in Agriculture: A Critical Appraisal of Genetic Engineering." *BioScience* 44(10): 677-689.

———. 1994b. "Using Hierarchy Theory to Explore the Concept of Sustainable Development." *Futures* 26(6): 616-625.

———. In press. "Socioeconomic Pressure, Demographic Pressure, Environmental Loading and Technological Changes in Agriculture." *Agriculture, Ecosystems and Environment*.

Giampietro, M., S.G.F. Bukkens, and D. Pimentel. 1993. "Labor Productivity: A Biophysical Definition and Assessment." *Human Ecology* 21(3): 229-260.

Giampietro, M., S.G.F. Bukkens, and D. Pimentel. 1997. "Linking Technology, Natural Resources, and the Socioeconomic Structure of Human Society: Examples and Applications. Pp. 131-200 in *Advances in Human Ecology*, Vol. 6, edited by L. Freese. Greenwich, CT: JAI Press.

Giampietro, M., G. Cerretelli, and D. Pimentel 1992a. "Energy Analysis of Agricultural Ecosystem Management: Human Return and Sustainability." *Agriculture, Ecosystems and Environment* 38: 219-244.

Giampietro, M., G. Cerretelli, and D. Pimentel. 1992b. "Assessment of Different Agricultural Production Practices." *AMBIO* 21(7): 451-459.

Giampietro, M., and D. Pimentel. 1991a. "Energy Efficiency: Assessing the Interaction Between Humans and Their Environment." *Ecological Economics* 4: 117-144.

Giampietro, M., and D. Pimentel. 1991b. "Energy Analysis Models to Study the Biophysical Limits for Human Exploitation of Natural Processes." Pp. 139-184 in *Ecological Physical Chemistry*, edited by C. Rossi and E. Tiezzi. Amsterdam: Elsevier Science Publishers B.V.

Giampietro, M., and D. Pimentel 1992. "Energy Efficiency and Nutrition in Societies Based on Human Labor." *Ecology of Food and Nutrition* 28: 11-32.

Gilliland, M.W. 1975. "Energy Analysis and Public Policy." *Science* 189: 1051-1056.

Hall, C.A.S., C.J. Cleveland, and R. Kaufmann. 1986. *Energy and Resource Quality*. New York: John Wiley & Sons.

Ho, M.W., and P.T. Saunders, eds. 1984. *Beyond Neo-Darwinism: An introduction to the New Evolutionary Paradigm*. London: Academic.

Jørgensen, S.E. 1992. *Integration of Ecosystem Theories: A Pattern*. Dordrecht: Kluwer.

Koestler, A. 1969. "Beyond Atomism and Holism: The Concept of the Holon." Pp. 192-232 in *Beyond Reductionism*, edited by A. Koestler and J.R. Smythies. London: Hutchinson.

Layzer, D. 1988. "Growth of Order in the Universe." Pp. 23-40 in *Entropy, Information, and Evolution*, edited by B.H. Weber, D.J. Depew and J.D. Smith. Cambridge, MA: MIT Press.

_____. 1990. *Cosmogenesis: The Growth of Order in the Universe*. New York: Oxford University Press.

Lotka, A.J. 1956. *Elements of Mathematical Biology*. New York: Dover Publications.

Mayumi, K. 1991. "Temporary Emancipation from Land: From the Industrial Revolution to the Present Time." *Ecological Economics* 4: 35-56

Morowitz, H.J. 1979. *Energy Flow in Biology*. Woodbridge, CT: Ox Bow Press.

Murphy, M.P., and L.A.J. O'Neil, eds., 1994. *What Is Life: The Next Fifty Years. Reflections on the Future of Biology*. Cambridge: Cambridge University Press.

Nicolis, J.S. 1986. *Dynamics of Hierarchical Systems*. New York: Springer-Verlag.

O' Neill, R.V. 1989. "Perspectives in Hierarchy and Scale." Pp. 140-156 in *Perspectives in Ecological Theory*, edited by J. Roughgarden, R.M. May and S. Levin. Princeton, NJ: Princeton University Press.

O'Neill, R.V., D.L. DeAngelis, J.B. Waide, and T.F.H. Allen. 1986. *A Hierarchical Concept of Ecosystems*. Princeton, NJ: Princeton University Press.

Odum, H.T. 1971. *Environment, Power, and Society*. New York: Wiley-Interscience.

_____. 1983. *Systems Ecology*. New York: John Wiley.

_____. 1996. *Environmental Accounting: Emergy and Decision Making*. Gainesville: University of Florida Press.

Odum, H.T., and R.C. Pinkerton. 1955. "Time's Speed Regulator: The Optimum Efficiency for Maximum Power Output in Physical and Biological Systems." *American Scientist* 43: 331-343.

Olsen, M.E. 1993a. "Components of Socioecological Organization: Tools, Resources, Energy, and Power." Pp. 35-67 in *Advances in Human Ecology*, Vol. 2, edited by L. Freese. Greenwich, CT: JAI Press.

Olsen, M.E. 1993b. "A Socioecological Perspective on Social Evolution." Pp. 69-92 in *Advances in Human Ecology*, Vol. 2, edited by L. Freese. Greenwich, CT: JAI Press.

Pastore, G., M. Giampietro, and K. Mayumi. 1996. "'Bioeconomic Pressure' as Indicator of Material Standard of Living." Paper presented at the fourth Biennial Meeting of the International Society for Ecological Economics (ISEE), August 4-7, Boston University.

Pattee, H.H., ed. 1973. *Hierarchy Theory: The Challenge of Complex Systems*. New York: Braziller.

Pillet, G., A. Baranzini, M. Villet, and G. Collaud. 1987. "Exergy, Emergy, and Entropy." Pp. 277-302 in *Environmental Economics: The Analysis of a Major Interface*, edited by G. Pillet and T. Murota. Geneva: Roland Leimgruber.

Pimentel, D., and M. Pimentel. 1979. *Food, Energy, and Society*. London: Edward Arnold.

Prigogine, I. 1978. *From Being to Becoming*. San Francisco: W.H. Freeman.

Rees, W.E., and M. Wackernagel. 1994. "Ecological Footprints and Appropriated Carrying Capacity: Measuring the Natural Capital Requirements of the Human Economy." Pp. 362-390 in *Investing in Natural Capital*, edited by A.M. Jansson, M. Hammer, C. Folke, and R. Costanza. Washington, DC: Island Press.

Rosen, R. 1985. *Anticipatory Systems: Philosophical, Mathematical and Methodological Foundations.* New York: Pergamon.

_____. 1991. *Life Itself: A Comprehensive Inquiry into Nature, Origin and Fabrication of Life.* New York: Columbia University Press

Salthe, S.N. 1985. *Evolving Hierarchical Systems: Their Structure and Representation.* New York: Columbia University Press.

_____. 1993. *Development and Evolution: Complexity and Change in Biology.* Cambridge, MA: The MIT Press.

Schneider, E.D., and J.J. Kay 1994. "Life as a Manifestation of the Second Law of Thermodynamics." *Mathmatical Computer Modelling* 19: 25-48.

Simon, H.A. 1962. "The Architecture of Complexity." *Proceedings of the American Philosophical Society* 106: 467-482.

Slesser, M. 1978. *Energy in the Economy.* New York: St. Martin's Press.

Smil, V. 1991. *General Energetics: Energy in the Biosphere and Civilization.* New York: John Wiley & Sons.

Tainter, J.A. 1988. *The Collapse of Complex Societies.* Cambridge: Cambridge University Press.

Ulanowicz, R.E. 1986. *Growth and Development: Ecosystem Phenomenology.* New York: Springer-Verlag.

Watt, K. 1989. "Evidence for the Role of Energy Resources in Producing Long Waves in the United States Economy." *Ecological Economics* 1: 181-195

Watt, K. 1992. *Taming the Future.* Davis, CA: Contextured Web Press.

Weber, B.H., D.J. Depew, C. Dyke, S.N. Salthe, E.D. Schneider, R.E. Ulanowicz, and J.S. Wicken. 1989. "Evolution in Thermodynamic Perspective: An Ecological Approach." *Biology and Philosophy* 4: 373-405.

Weber, B.H., D.J. Depew, and J.D. Smith, eds. 1988. *Entropy, Information, and Evolution: New Perspectives on Physical and Biological Evolution.* Cambridge, MA: MIT Press.

White, L.A. 1943. "Energy and the Evolution of Culture." *American Anthropologist* 14: 335-356.

White, L.A. 1959. *The Evolution of Culture: The Development of Civilization to the Fall of Rome.* New York: McGraw-Hill.

Whyte, L.L., A.G. Wilson, and D. Wilson, eds. 1969. *Hierarchical Structures.* New York: Elsevier.

Wicken, J.S. 1987. *Evolution, Information and Thermodynamics: Extending the Darwinian Program.* Oxford: Oxford University Press.

Zipf, G.K. 1941. *National Unity and Disunity: The Nation as a Bio-social Organism.* Bloomington, IN: The Principia Press.

LINKING TECHNOLOGY, NATURAL RESOURCES, AND THE SOCIOECONOMIC STRUCTURE OF HUMAN SOCIETY:
EXAMPLES AND APPLICATIONS

Mario Giampietro, Sandra G.F. Bukkens,
and David Pimentel

ABSTRACT

The preceding paper proposed to study changes in socioeconomic systems based on the congruence between profiles of the requirement and of the supply of exosomatic energy and human time, measured over different socioeconomic categories. Here, we present six applications: (1) a definition and validation of a set of biophysical indicators of human development that are able to bridge different readings (social, economic, etc.) of such a process; (2) the use of dynamic energy budget analysis to study the relations between

---

Advances in Human Ecology, Volume 6, pages 131-200.
ISBN: 0-7623-0257-7

technological development and demographic changes at the level of the socioeconomic system; (3) a proposal for operational definitions of carrying capacity and optimum population; (4) a modeling of pre-industrial and industrial societies as dynamical systems; (5) a critical examination of "envisioning," by checking future scenarios of socioeconomic development against biophysical constraints; and (6) an exploration of the mechanism that tends to drive human societies toward unsustainability.

BIOPHYSICAL INDICATORS OF DEVELOPMENT

Development of society implies a change in the structure and functions of the socioeconomic system. Given the complex nature of any socioeconomic system, development can be described and analyzed from several different perspectives (based on descriptions referring to different spatiotemporal scales), such as individual human beings, local communities, the economy, and the natural environment. A comprehensive assessment of societal development not only requires the use of all these different windows of observation. It also requires linking and weighing the different conceptualizations of development thus obtained. Biophysical analyses of socioeconomic systems are useful to bridge the different readings of development by enabling one to assess a congruence of events taking place in parallel on the different levels of the system. For example, an input/output analysis of the economic process can establish a relation between environmental loading, the "outcome" of development as seen from the ecological perspective, and material standard of living, the "outcome" of development as seen from the point of view of the economist.

In the preceding paper (Giampietro 1997, this volume), a theoretical link was established between technology, natural resources, and socioeconomic structure by analyzing the dynamic nature of the energy budget of society. Societal development was described in terms of changes in the patterns of allocation of useful energy and human time. It was shown that development implies the ability of the socioeconomic system to reach a high exosomatic/endosomatic energy ratio as well as the ability to allocate an increasing fraction of the available human time to "adaptability" (a high value for the ratio between non-working time and total human time).

Following this rationale, we here propose a biophysical indicator of socioeconomic development as perceived by individuals in society. The indicator was briefly introduced in the preceding paper and referred to as bioeconomic pressure, or BEP. Bioeconomic pressure applies to the society and can be considered a biophysical substitute for economic indicators such as the gross national product (GNP). Being defined at the level of society as a whole, BEP does not address the issue of equity.

The equation defining BEP was:

$$BEP = ET/C = (ABM \times MF) \times (Exo/Endo) \times (THT/C) \qquad (1)$$

Variables Defining BEP: Meaning and Range of Values

The variables defining the bioeconomic pressure in society are shown on the right side of Equation (1). They can be considered as three separate terms (as indicated by the parentheses), each of which reflects a different view on the material standard of living in society and refers to a different hierarchical level of analysis. The values of each of these three terms as well as that of BEP have been calculated for a sample of 107 countries representing more than 90 percent of world population (Pastore et al. 1996). The main findings are summarized below.

The first term (MF × ABM), where MF stands for metabolic flow and ABM for average body mass of the population, reflects the per capita endosomatic energy metabolism in society (MJ of food energy/hour). Hence, this term applies to individuals. The higher this value, the better are the physiological conditions of the humans living in society. For the examined sample of countries, the values of this term ranged between a minimum value of 0.33 MJ/hr, typical of the poorest developing countries, and a maximum value of 0.41 MJ/hr, typical of developed countries where the value reaches a plateau. The range is very narrow, and this is due to the fact that the two factors MF and ABM are inversely correlated to development. ABM increases from less than 35 kg in subsistence societies up to almost 60 kg in developed societies, whereas MF decreases from almost 9.5 kJ/kg/hour in subsistence societies to less than 7 kJ/kg/hour in developed societies (see Table 1 in Giampietro 1997, this volume; for more details, see Giampietro et al. 1993).

The second term, "exo/endo energy ratio," is a measure of the exosomatic energy metabolism in society. This term applies to the socioeconomic system as a whole. Note that the exosomatic metabolism is expressed on a per capita basis as a multiple of the endosomatic metabolism (ABM x MF). Because the range of values of the endosomatic metabolism is very narrow (see above), variations in the exo/endo ratio basically reflect differences in the exosomatic metabolism of society. For most of the countries examined, this term ranged from a minimum value of 5 (countries where exosomatic energy is basically in the form of traditional biomass fuels and animal power) to a maximum value of 90 (countries where exosomatic energy is basically in the form of machine power and electricity obtained by relying on the depletion of fossil energy stocks). The exo/endo ratio is a good indicator of economic activity, for it is strongly correlated to the GNP per capita. The higher the exo/endo ratio, the more goods and services that are produced and consumed per capita in society.

The third term (THT/C), where THT stands for total human time available in society and C for time allocated to work in the primary sectors of the economy, also refers to the socioeconomic system as a whole. It reflects the relative amount of human time that is allocated to activities that provide for long-term adaptability of the system. The minimum value is around 10, found in subsistence systems where agriculture absorbs a large fraction of the work force; and the maximum value is above 40, found in post-industrial societies with a large fraction of the work force engaged in the service sector and a large part of the population not economically active. This indicator reflects the implications of "development" on the social organization of society (length of education, retirement, leisure time for workers, and size of service sector, including health care). Note that a certain bias in the values is present because the definition of A, B, and C activities adopted in the model is based on Western socioeconomic systems. In subsistence societies, very often the distinction between these activities is less clear (see Giampietro et al. 1993).

The bioeconomic pressure—BEP, the combination of the above three terms—provides an overall assessment of the material standard of living in society taking into account several perspectives: the physiological/individual (MF × ABM), the economic/technological (exo/endo ratio), and the social (THT/C). BEP increases with any change that improves the material standard of living in society. Its minimum value can be lower than 15 MJ/hr in the poorest developing countries, and its maximum value can be as high as 1500 MJ/hr in developed countries.

Correlation of Bioeconomic Pressure with Classic Indicators of Development

The use of BEP—and that of each of its three composite terms—as indicators of human development has been tested by comparing their value to 24 classic indicators of development (those used in the World Bank tables) over a sample of 107 countries (Pastore et al. 1996). The analysis shows

Table 1.   Correlation (r) Between BEP, Exo/Endo Energy Ratio, THT/C, (ABM × MF), and Some Indicators of Nutritional Status and Physiological Well-being

	Log (BEP)	Log(Exo/Endo)	THT/C	ABM × MF
Life expectancy	0.79	0.75	0.63	0.59
Energy intake	0.82	0.81	0.55	0.73
Fat intake	0.87	0.85	0.63	0.77
Protein intake	0.85	0.85	0.57	0.72
Child malnutrition	−0.71	−0.65	−0.63	−0.70
Infant mortality	−0.76	−0.74	−0.57	−0.58
Low birth weight	−0.65	−0.62	−0.49	−0.63

Table 2.   Correlation (r) Between BEP, Exo/Endo Energy Ratio, THT/C, (ABM × MF), and Some Indicators of Economic and Technological Development

	Log (BEP)	Log(Exo/Endo)	THT/C	ABM × MF
Log (GNP)	0.92	0.89	0.63	0.66
% GDP produced in agriculture	−0.77	−0.73	−0.60	−0.54
Log (GNP/hour labor)	0.92	0.87	0.71	0.63
% labor force in agriculture	−0.90	−0.81	−0.72	−0.66
% labor force in services	0.90	0.83	0.76	0.56
Energy consumption/capita	0.92	0.95	0.53	0.67
Expenditure for food	−0.86	−0.87	−0.69	−0.78

that both the overall index BEP as well as each of the three separate terms are strongly correlated with the traditional indicators of development found in world statistics. Examples of graphs reporting this correlation are presented in Figures 2 and 3 of the preceding chapter (Giampietro 1997, this volume). An overview of all correlation factors is given here in Tables 1-3.

In the cross-sectional analysis of major countries, BEP shows the same correlation with classic indicators of development obtained when using GNP (Pastore et al. 1996). An important conclusion from the correlation analysis is that, compared to the widely used economic indicator GNP, the biophysical indicator BEP offers three main advantages: (1) BEP can also be applied in nonmonetarized societies, whereas the GNP loses its coherence in these societies; (2) BEP can be calculated at the level of villages, cities, or particular social groups; and (3) BEP refers to actual changes in biophysical parameters that determine material standard of living and, unlike the GNP, is not affected by inflation (longitudinal studies) or exchange rates among currencies. Figure 1 presents data from a longitudinal study of OECD countries in the period 1978-1992. It shows that assessments given in brute GNP are subject to fluctuations due to inflation and

Table 3.   Correlaton (r) Between BEP, Exo/Endo Energy Ratio, THT/C, (ABM × MF), and Some Indicators of Social Development

	Log (BEP)	Log(Exo/Endo)	THT/C	ABM × MF
Televisions/inhabitant	0.89	0.89	0.62	0.72
Cars/inhabitant	0.88	0.91	0.59	0.72
Newspapers/inhabitant	0.77	0.80	0.47	0.60
Phones/inhabitant	0.87	0.88	0.61	0.71
Log (population/physicians)	−0.81	−0.76	−0.60	−0.67
Log (population/hospital beds)	−0.77	−0.78	−0.51	−0.70
Pupil/teacher ratio	−0.77	−0.76	−0.51	−0.62
Illiteracy rate	−0.61	−0.58	−0.42	−0.44
Primary school enrollment	0.44	0.39	0.38	0.36
Access to safe water	0.78	0.77	0.53	0.59

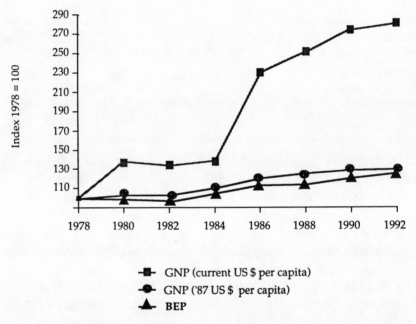

-■- GNP (current US $ per capita)
-●- GNP ('87 US $ per capita)
-▲- BEP

Figure 1.   BEP and GNP per capita of OECD countries in the period 1978-1992.

changing exchange rates among currencies (Pastore et al. 1996). For example, the sharp devaluation of the U.S. dollar in the 1980s generated a "virtual" increase in the GNP of OECD countries. A correct economic reading of longitudinal trends and cross-sectional comparisons in terms of GNP requires corrections for inflation and fluctuations in exchange rates, whereas indications provided by BEP are straightforward and similar to those given by the corrected GNP.

LINK BETWEEN TECHNOLOGICAL DEVELOPMENT AND DEMOGRAPHIC CHANGES

Development and Demographic Transition in Terms of the Model

The development of socioeconomic systems can be seen as an intensification of the informed autocatalytic loop that stabilizes the dynamic equilibrium between the supply and the requirement of useful energy in society. This process can be conveniently described by a combination of intensive and extensive variables. The former are related to the equilibrium between bioeconomic pressure (BEP) and strength of the hypercycle (SEH), discussed in detail in the preceding paper (Giampietro 1997, this volume). The latter include population size (endosomatic energy flow) and total energy throughput (exosomatic energy flow).

The use of both intensive and extensive variables helps to distinguish the *growth* and *development* of socioeconomic systems. Growth refers to an increase in size—that is, "having more of the same thing"—and is indicated by an increase in total energy throughput (ET) in society at constant bioeconomic pressure (BEP). Development refers to a change in internal characteristics of society and is reflected by an increase in bioeconomic pressure (BEP). Note that both growth and development affect the stability of boundary conditions (environmental loading).

The development of society is intricately linked to changes in population size and structure. In terms of our model, the classic demographic transition can be seen as a movement from one metastable equilibrium of the energy budget to another, involving three phases. In the first phase, society finds itself in a metastable equilibrium in which the energy requirement matches the energy supply under conditions of low labor productivity (SEH) and low bioeconomic pressure (BEP). This situation of no growth and no development is followed by a second, transitional phase that is triggered by an increase in strength of the autocatalytic loop of exosomatic energy. In this transitional phase, the consequent increase of the average energy return of labor (SEH > BEP) generates an energy supply that exceeds the requirement. The energy surplus is initially absorbed by an increase in population size (extensive variable), while the bioeconomic pressure (intensive variable) remains close to the original value. Hence, at the beginning this transitional phase is characterized by pure growth. After a certain lag time, however, the bioeconomic pressure also increases. This results in both growth and development, the latter of which will dominate toward the end of the transitional phase. A new metastable equilibrium (third phase) will finally be reached after a certain time, when the values of the variables determining the bioeconomic pressure have "settled." In the period of settlement, the new (more and better) exosomatic devices and new (more and better) controls on the autocatalytic loop of exosomatic energy imply relevant changes in the cultural identity of the system. The results are to create and stabilize new patterns of human time allocation (process of development).

The Demographic Transition in Equations

In the preceding paper, the following equation was introduced. This equation defines an equilibrium between energy requirement (BEP, left side of the equation) and energy supply (SEH, right side of the equation) in society:

$$(ABM \times MF) \times Exo/Endo \times THT/C = ET/CI \times CI/C \qquad (2)$$

The terms on the left side of Equation (2) define the bioeconomic pressure and refer to collective properties of the socioeconomic system. Two of these terms affect and are affected by demographic changes, namely, (ABM × MF) and THT/C. The two terms on the right side of the equation are linked to technological development and the availability of natural resources and thus are scale-dependent and affected by population size (Giampietro and Mayumi 1996).

To effectively describe demographic changes in terms of energy analysis, we must reformulate the equation of balance, Equation (2), so as to include *extensive* variables. To this purpose, we express the total energy throughput (ET) in society, which is linked to population size, as a function of the terms defining SEH and BEP:

$$ET_i = exo/endo_i \times POP_i \times (ABM \times MF)_i \qquad (3)$$

When the supply of useful energy (SEH) exceeds the requirement (BEP), socioeconomic systems will adjust their parameters so as to augment the use of (surplus) useful energy, thereby expanding the scale of their activity (resulting in larger ET). Thus, at a point in time $i+1$, we should then have:

$$ET_{i+1} > ET_i \qquad (4)$$

and thus:

$$(exo/endo)_{i+1} \times (pop)_{i+1} \times (ABM \times MF)_{i+1}$$
$$> (exo/endo)_i \times (pop)_i \times (ABM \times MF)_i \qquad (5)$$

This equation shows that an increase in energy dissipation ($ET_{i+1} > ET_i$) for the socioeconomic system can be obtained through an increase in endosomatic metabolism of the population ($ABM \times MF$), an increase in population size (pop), or an expansion of exosomatic devices operating in the socioeconomic system (exo/endo). The range of change of the first term is very limited ($ABM \times MF$ can only move from 0.33 to 0.41 MJ/hour/capita). Expressing the ratio of exosomatic and endosomatic energy as:

$$Exo/Endo = ET/CI \times CI/C \times C/THT \times 1/(ABM \times MF), \qquad (6)$$

equation (5) can also be formulated as:

$$(ET/C)_{i+1} \times (C/THT)_{i+1} \times (pop)_{i+1} > (ET/C)_i \times (C/THT)_i \times (pop)_i \qquad (7)$$

Equation (7) gives a better view on the possible ways of expanding ET. When the energy surplus is absorbed by an increase in the exo/endo ratio rather than by an increase in population size, the solution implies an increase in ET/C, that is, the bioeconomic pressure BEP [see Equation (1)]. At fixed population size, increases in ET/C are translated into a decrease of C/THT (increased allocation of human time to adaptability). Put another way, depending on the path of expansion followed by the system, increases in ET can be transformed into either increased population size or improved material standard of living. The latter solution implies a growing investment of the energy throughput and human capability for providing control in adaptability.

Changes in ET/C and in THT/C are subject to biophysical and cultural constraints that create certain "intrinsic" lag times in the adjustment of their values to new conditions. Notably, they depend on the rate of industrialization (e.g., availability of technological capital to alter the nature of the hypercycle) and the rate at which the pattern of human time allocation to societal activities is changed (e.g., cultural resistance to changing the variables determining the bioeconomic pressure). The difference in lag time between changes in ET/C and changes in THT/C will determine how much of the increased energy throughput will be expressed as population growth and how much as improved material standard of living (the shape of the flex in the sigmoid curve representing the demographic transition in time).

Numerical Examples

To evaluate possible combinations of values for the equilibrium indicated in Equation (5), we first reformulate (for the sake of clarity) this equation as:

$$ET/capital/year = (ET/CI \times CI/C) \times C/THT \qquad (8)$$

Data from our analysis at the country level suggest that two different solutions are possible for this equation.

Type 1 Equilibrium

This solution is represented by pre-industrial societies. Societies with this equilibrium have a low energy consumption (ET) per capita, in the order of 6-10 GJ/year, and they rely on an exosomatic autocatalytic loop that has a ratio ET/CI greater than 3. These societies typically have a low ratio of CI/C (3-5 MJ/h), for they make little use of machine power (muscle power per worker is about 0.1 HP). Their ratio of C/THT is relatively high (0.08-0.10, or THT/C equal to 10-12) when compared to developed societies. Hence, a large fraction of their labor force is engaged in agriculture. Demand and supply in the dynamic energy budget are matched and stabilized at a poor material standard of living (BEP = SEH < 50 MJ/hour). The population size generally is small, in the order of magnitude of hundreds of people.

Type 2 Equilibrium

This solution is represented by post-industrial societies. The solution implies a high energy consumption (ET) per capita of about 100-300 GJ/year, obtained through the use of an exosomatic autocatalytic loop with a low ratio ET/CI (around 2). The ratio of CI/C reaches very high values (200-500 MJ/hour) and indicates total reliance on machine power, with a power level per worker of over hundreds of HP. The ratio of C/THT is relatively low (0.03, or THT/C ≈ 40) compared to pre-industrial societies. Only a small fraction of the labor force is engaged in agriculture and mining. Demand and supply in the dynamic energy budget are stabilized at a high material standard of living (typical values are BEP = SEH > 500 MJ/ hour). The size of these systems is typically large, in the order of magnitude of hundreds of millions of people. Therefore, a high level of consumption per capita implies the risk of heavy environmental loading depending on the size of the ecosystems supporting societal activity.

Transition Between Two Metastable Equilibria: The Industrial Revolution

Imagine a process of expansion of ET starting from a pre-industrial socioeconomic system in phase 1 of the demographic transition. Hence, we start from a situation where the initial population size is small and stable, although subject to large fluctuations caused by high fertility and mortality rates. Assume now that improvements in the exosomatic autocatalytic loop (e.g., new technologies, use of higher-quality resources, and better knowledge and management of socioeconomic activities) make a surplus of energy available to the socioeconomic system (SEH>BEP). The surplus can be absorbed by an increase in population or by an increase in energy consumption per capita, either of which will be reflected in a changing pattern of human activity (changes in THT/C). In the short term, population growth inevitably generates an increase in THT/C because of the increasing number of children in the system. On a longer time scale, the bigger size of the socioeconomic system due to population growth will cause a further increase in THT/C because of an increasing ratio of B/ (B+C). Indeed, with the expanding size of society, the fraction of total work supply allocated to administration and other services gradually has to increase. However, in pre-industrial societies the possibility of increasing the ratio THT/C is limited due to the lack of devices that could amplify the power of workers in the primary sectors (ET/C in pre-industrial societies is small).

History shows that wind and water power were decisive factors in determining local spots of development in pre-industrial societies. Nevertheless, the very nature of these "power devices"—they are location-specific, or do not provide a continuous (reliable) flow of power in time—prevented pre-industrial socioeconomic systems from reaching a type 2 equilibrium on a large scale. The only solution for pre-industrial societies to reach a large-scale, complex organization was the division of the population into widely different social classes and a heavy taxation of farmers by the central administration (to increase the ratio B/B+C). However, this could not avoid a strong instability of these systems in the long term (discussed below).

The introduction of fossil energy sources and thermodynamic converters (engines) radically changed the picture, as is indicated by the term industrial *revolution*. Fossil energy sources (initially coal) represented a virtually unlimited supply of energy input, while thermodynamic converters guaranteed a reliable and efficient way to convert this energy input into mechanical power (avoiding bottlenecks in the investment of useful energy in the CI sector). For the first time, the ratio of CI/C became a variable subject to human control. It could be augmented by increasing the investments in technological development, albeit at the cost of a decrease in the ratio of ET/CI. However, whenever surplus ET was available, new paths of expansion of the socioeconomic system were opened.

In fact, in pre-industrial societies, where biomass was the main energy input and human and animal muscle the main source of power, a higher energy dissipation could only be achieved in society by exploiting a larger area of ecosystem per capita. This solution soon reached a limit—first, because the exploitation of larger areas required more labor per unit of area (e.g., transportation costs increased more than linearly) and hence sooner or later resulted in a decrease in SEH; and second, because such a solution was simply not feasible, especially in cases of excessive increase in population size when there were no new territories in which the excess of population could expand. In industrial times the picture has totally changed, and population growth no longer implies a reduction in SEH because the supply of energy input is no longer related to the availability of land, and surplus ET can be absorbed not only by population growth but also by an increase in the exo/endo ratio.

The dynamic transition from type 1 to type 2 equilibria according to Equation (8) is described in Figure 2. All variables in this figure are intensive, in order to avoid a complicated three-dimensional representation, but we can imagine a third axis perpendicular to the plane indicating the size (in terms of population) of the system.

Data (from the database presented in Pastore et al. [1996]) are used in Figure 3 to represent the same two equilibria indicated in Figure 2. The white and black dots in the two quadrants (upper-right and lower-left) of Figure 3 derive from a synchronic analysis of different countries (where white dots = 107 countries aggregated in clusters, with values referring to the year 1992) and a diachronic analysis (where black dots = average values referring to OECD countries in 1970, 1980, 1985, 1990, and 1993). It

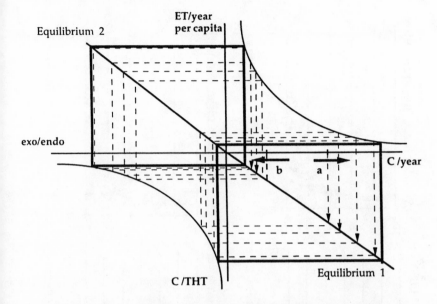

Figure 2.   Demographic transition as a shift between two equilibria
of the energy budget of society.

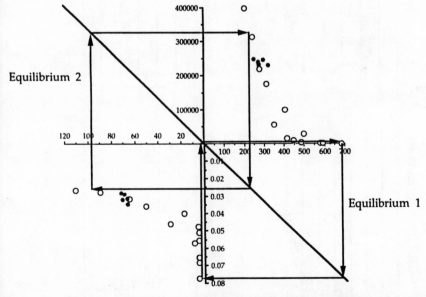

Figure 3.   The two equilibria of the demographic transition based on actual data
from a synchronic (107 countries in 1992) and a diachronic (OECD countries over the
last 23 years) analysis.

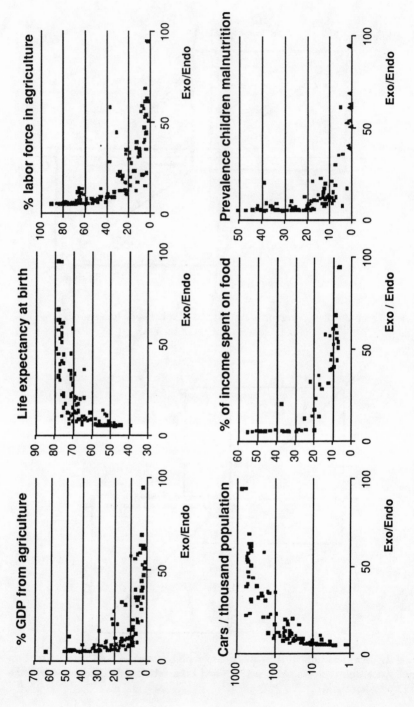

Figure 4. Correlation of exo/endo energy ratio with major indicators of development.

is very interesting to note that the black and white dots are on the same curves.

Pastore et al. (1996) have plotted for the same sample of countries the exo/endo energy ratio against the 24 indicators of development referred to earlier. An example is given in Figure 4. With these countries being in different phases of the transition between the two metastable equilibria, this cross-sectional picture illustrates the trajectory followed in the transition. Socioeconomic systems appear to reach type 2 equilibrium after the exo/endo ratio passes a threshold value of about 30. A similar trajectory was found (for the same sample of countries) for the other two terms making up BEP: The type 2 equilibrium is reached by countries that have surpassed the threshold values of about 9 MJ per capita per day for (ABM × MF), and of about 40 for the ratio THT/C (Pastore et al. 1996).

In conclusion, these data show that the classic description of the demographic transition in terms of indicators of fertility and mortality should thus be seen as just one of the possible descriptions of the process of transition between two metastable equilibria of the dynamic energy budget. Biophysical indicators such as those presented here can equally well describe and explain the transition process.

CARRYING CAPACITY AND OPTIMUM POPULATION

The model presented in the preceding paper (Giampietro 1997, this volume) provides useful starting points to operationalize the concepts of *carrying capacity* and *optimum population*. Particularly relevant in this context are the scale-dependent parameters CI/C (productivity of labor in the primary sectors) and ET/CI (efficiency of the exosomatic energy loop) that characterize the autocatalytic loop of exosomatic energy in society. Given the effects of economies of scale and specialization, it can be assumed that CI/C tends to increase with the size of the system. Initially, ET/CI is likely to be positively correlated with the size of society but, assuming fixed natural capital, it will eventually decrease with a further enlargement of the system. This decreasing marginal return of the energetic investment in CI activities is due to two causes: (1) The growth in the size of the throughput implies a rapid decrease in the quality of resources (the better ones are exploited first). This presents the problem of finding substitutes for limited resources at an increasing speed. (2) Ecological problems increase when the environmental loading exceeds the environmental sink capacity.

The combination of these contrasting effects of scale is useful to explore the concepts of carrying capacity and optimum population (Giampietro and Mayumi 1996).

Human Society on a Space BEP-EL-SIZE

Our model of analysis relates the bioeconomic pressure in society (BEP) and the environmental loading (EL) through intensive and extensive variables. Therefore, the sustainability of a defined energy budget depends on characteristics of the society (e.g., cultural identity, laws, and technology) and the characteristics of the ecosystem being exploited (e.g., size, typology of biological communities, and climate). Although the characteristics of the ecosystem and the society will change in the long run, on a short time scale, energy flows can be assumed to be in quasi-steady state. Then, we can assume that the particular process of self-organization of society is described by a particular combination of bioeconomic pressure (BEP), environmental loading (EL), and size of society (SIZE).

Comparing societies that operate at different values of BEP and EL by constructing a space, BEP-EL-SIZE, puts the technological development of society in perspective. For example, the socioeconomic system of the Yanomamö and of Japan are illustrated on the space BEP-EL-SIZE in Figure 5. The stability of the equilibrium between BEP and SEH is subject to internal and external constraints. Concerning the former, to be sustainable the process of self-organization must result in a BEP that is socially acceptable and at the same time be based on a SEH that is biophysically feasible (that is, on the availability of exosomatic devices and a system of controls to stabilize the hypercycle). Concerning the external constraints, to be stable, the equilibrium between BEP and SEH must be ecologically compatible. Numerical values for the axis BEP = SEH in Figure 5 are easily calculated, whereas the choice and assessment of a set of indicators of environmental loading (EL) is a more delicate task.

In fact, it is quite possible to obtain different EL indicators for the same system. For example, an environmental loading in terms of energy controlled by humans versus energy controlled by the ecosystem can be measured in several different ways: by assessing actual flows of energy consumed in a defined area (e.g., country surface), by assessing embodied quantities of energy needed to stabilize flows of energy consumed in an area, by transforming energy flows obtained from depletion of stocks into an "energy flow equivalent" and calculating the size of the ecosystem that

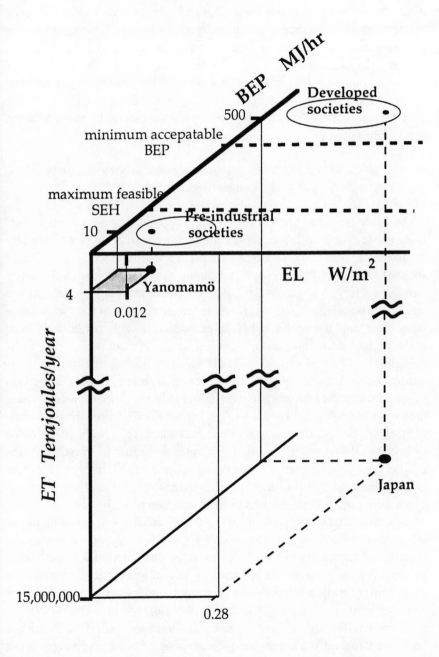

Figure 5. Socioeconomic systems described on a space: BEP-EL-SIZE.

would be needed to generate them, and so forth. Any one of these choices defines a distinct EL and scale of analysis. In Figure 5, we opted for an assessment of the exosomatic energy consumed per square meter of the area actually occupied by society ($Watt/m^2$). Clearly, a different choice of EL would put the focus on different aspects of the process of self-organization (e.g., embodied energy used by Japan in its imports). Obviously, once it is chosen, we have to consistently use the same EL indicator when comparing different socioeconomic systems.

Using the Space BEP-EL-SIZE to Define Carrying Capacity and Optimum Population

The space BEP-EL-SIZE is useful in conceptualizing definitions of carrying capacity and optimum population within the framework provided by the model. Figure 6 shows a plane BEP-EL for different values of size of society. On a short time scale, assuming flows of matter and energy through society in a steady state, we can define carrying capacity as the maximum population size for which an equilibrium between the requirement (BEP) and the supply (SEH) of exosomatic energy per hour of work in the primary economic sectors is still possible, at a defined level of technological and natural capital, with the value of BEP not below the minimum acceptable to the people in the society. Increases in population size beyond the carrying capacity will result in too high an environmental loading or too low a standard of living. Similarly, we can define optimum population as the population size that, according to a defined level of technological and natural capital, generates a maximum in SEH (BEP). Further population growth beyond the optimum size will imply an increase in environmental loading that will negatively affect the strength of the hypercycle (causing a reduction in SEH and then in BEP).

To be meaningful, the concepts of carrying capacity and optimum population must be defined on a time scale shorter than the one at which the society and ecosystem evolve. Otherwise, the characteristics of both society and ecosystem can not be assumed as "given" at a particular moment in time; consequently, a relation between population mass and SEH (through an increase in efficiency and economies of scale), and between population mass and EL (through an increase in ET and a reduction in ET/CI), can not be defined. Indeed, if a longer time scale is adopted at which changes occur in the characteristics of society (e.g., technical coefficients, system of values, optimizing strategies) and ecosystem (e.g., stocks, sinks, and bio-

BEP = SEH

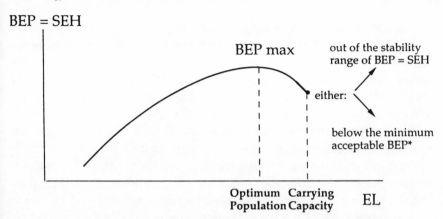

Figure 6.   Definition of Optimum Population and Carrying Capacity
on a plane BEP-EL.

geochemical cycles), the concepts of optimum population and carrying capacity have no meaning.

Furthermore, carrying capacity and optimum population can not be defined on the basis of only the characteristics of the society or the characteristics of the ecosystem. It is the *interaction* between society and ecosystem that determines the stability of the attractor BEP = SEH (examples are given in the simulation of pre-industrial societies below).

The concepts of carrying capacity and optimum population have to do with the contrasting effects that the scale of the socioeconomic system has on internal and external constraints. A larger scale of society is a positive factor with regard to internal constraints because it increases CI/C and enables society to generate more information related to adaptability. That is, a larger scale eases internal constraints, thus making possible higher values of BEP. On the other hand, a larger scale of the economy tends to create negative effects in terms of ET/CI because of the decreasing quality of natural resources (Hall et al. 1986), thereby increasing energy costs for food production (Pimentel and Pimentel 1979; Hall et al. 1986; Gever et al. 1991; Giampietro in press) and, sooner or later, increasing energy costs for pollution control. In fact, when the environmental loading becomes too heavy, then pollution control, monitoring, and the restoration of environmental equilibria should be added to the activities defined as the direct investment (DI) of society. This would reduce the strength of the hypercycle.

The two contrasting "scale effects" are more or less relevant in determining strategies of development, depending on the "openness" of the society (the extent of reliance on trade and stock depletion). The more the economy is open, the more it is tempting to neglect negative effects induced by a higher environmental loading and to focus on the positive aspects of a larger size. In fact, more developed societies can import ecosystem activity from distant ecosystems and export ecological problems to less developed societies by trading.

While it is easy to define carrying capacity and optimum population in theoretical terms, the operational definition of these concepts is based on problematic assessments of minimum acceptable BEP (social compatibility) and maximum acceptable EL (ecological compatibility). These assessments are observer-dependent (e.g., there are always social groups that do not like present conditions) and scale-dependent (nothing that self-organizes is stable in time). What is a reasonable definition of minimum acceptable BEP and maximum acceptable level of environmental risks? Note that changes on the human side, such as the ability to achieve a higher BEP, are detectable on a short time scale and, therefore, easier to quantify than adverse impacts on the environment, such as the erosion of ecosystem resilience. These are detectable only on a much longer time scale (Giampietro 1994a). As a consequence, whenever possible, industrial society tends to ignore the "rights" of biophysical systems, that is, the intrinsic ethical dimension of economic activity.[1]

DYNAMIC MODELS OF SOCIOECONOMIC SYSTEMS

Pre-industrial Societies Based on Fund-Type Energy Sources

Characteristics of the Energy Budget

Fund-type energy resources (see Giampietro 1997, this volume, note 6) are defined in their flow dimension. That is, the rate of generation of energy input harvestable in time per unit of land is defined by the process generating the supply of gradients of free energy (e.g., by the level of primary productivity of the ecosystem used as the source of biomass). Therefore, to double the per capita energy throughput (ET) in pre-industrial societies at a given technological level, society has to double the area exploited per capita per day. The general scheme of such a dynamic budget can be expressed for pre-industrial societies as shown in Figure 7.

The demand of human work per unit of energy consumed by society is what determines the labor productivity (SEH). For societies exploiting fund-type energy sources, this value can be obtained by interpolating in

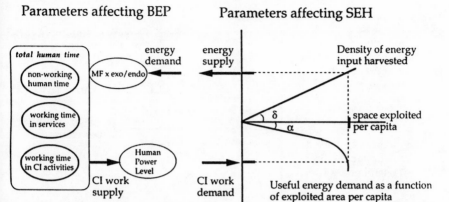

Figure 7. Parameters determining the energy budget based on "fund-type" (flow-limited) energy sources.

Figure 7, at a certain area of ecosystem exploited per capita: (1) the energy input harvested, which is proportional to the angle δ (the density of energy input, such as biomass, harvested by society); and (2) the demand of work, which is determined by the value assumed by the function "useful energy demand per area exploited" and by the quantity of power delivered per hour of labor, assessed by the human power level (HPL). The function of useful energy demand per exploited area starts with the angle α, which measures the direct investment (DI) of useful energy required to exploit the area in which energy input is collected. This function increases with the amount of area exploited in a nonlinear way because of the increasing costs of moving workers to and from the fields and of transporting materials.

Parameters affecting the ratio of energy demand per unit of work supply to the hypercycle are reported on the left side of Figure 7. The two parameters determining the allocation of human time (ratio of non-working/ working time and fraction of working time allocated to services) and those determining the requirement of energy input (metabolic flow and exo/endo energy ratio) depend on variables detectable at both the physiological and socioeconomic level.

Some characteristics of the hypercycle sustaining the process of energy dissipation in pre-industrial societies are typical for these societies. First, there is an upper limit to the net disposable energy income (the ratio FI/CI) that can be obtained by exploiting fund-type energy sources. After a certain threshold and at fixed technological capital, the cost of exploitation of increasing areas tends to increase more than the return it provides. In fact, a higher investment per hectare, in terms either of hours of labor or specific

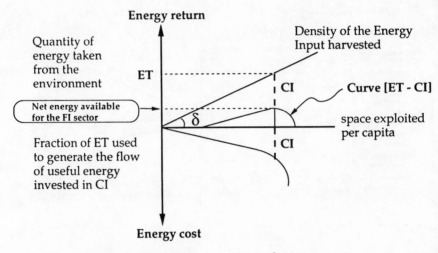

δ = Density of energy input used by society = e.g. W/m² of food, fuelwood;
ET = flow of Energy input harvested from the environment;
CI = Cost of Circulating Investment = the fraction of ET used to exploit the area
ET - CI = Fraction of ET that can be used to sustain activities in the FI sector

Figure 8. Hypercycle of a society exploiting "fund-type" (flow-limited) energy sources.

biophysical requirements of applied power, is inevitable because of increased travel and transportation. Thus, there is a decreasing marginal return for increasing the area exploited. This is portrayed in Figure 8.

Second, there is a strong internal constraint—portrayed in Figure 9—that limits the exo/endo energy ratio. This is related to the bottleneck in power level, because harvesting energy input over increasingly larger areas requires larger flows of applied power per hour of labor to reduce travel and transportation time and to manage a larger area per workday. Although the capacity of power delivery of human workers is extremely limited (about 0.1 HP), animal power does not dramatically change the picture. Animal power increases power delivery by a factor of 10 at best, and animal power demands space for feed production.

Third, there is an obvious external constraint on the hypercycle in the form of limited availability of natural capital per capita, notably land (or ecosystem area). When population growth reduces the area available per capita below the minimum required to provide the expected flow ET, the system encounters an "external limit" to its expansion, as seen in Figure 10, part a. The same applies when a too-intensive exploitation reduces the productivity of the ecosystem (through ecosystem degradation) on which

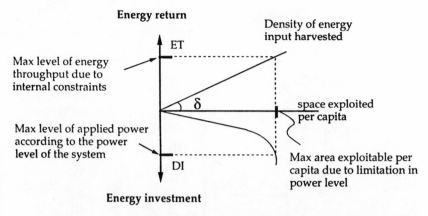

Figure 9. Internal constraint limiting the exploitation of "fund-type" (flow-limited) energy sources.

society depends for its energy input. Note the decrease from δ to δ* in Figure 10, part b.

It is possible to apply to these pre-industrial systems critical environmental loading ratios, as discussed in Giampietro (1997, this volume). They are to be considered as external constraints to the stability of the equilibrium of the energy budget (see again Figure 10, part a).

Biophysical Constraints Determined by the Energy Budget

Before presenting the simulation of pre-industrial societies with a dynamical model, let us explore in theoretical terms the nature of internal and external constraints affecting socioeconomic characteristics as determined by the need of balancing the energy budget. This theoretical discussion is based on several simplifications. First, the investment of useful energy in the CI sector is assumed to be proportional to the hours of human work delivered there (the exosomatic power delivered in the CI sector is assumed to be proportional to the working time of humans). Second, at fixed technological capital, we can define a ratio α of hours of labor required per hectare of a particular production system based on a defined set of activities (specific for the society under investigation). Third, the hours in a workday are defined (e.g., a maximum of 10 hours), and workers are assumed to travel from home to the exploited area and back once a day with a defined speed "ws". Last, the area exploited is circular around the village with radius r, and the village is supposed to occupy a negligible area in the center.

a. population density

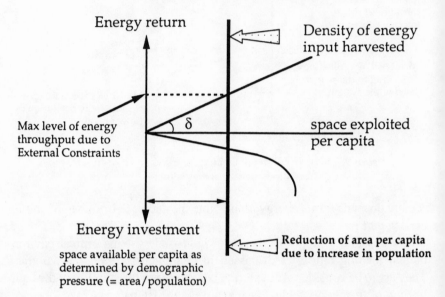

b. environmental degradation

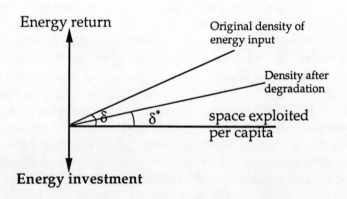

Figure 10.   External constraints limiting the exploitation of "fund-type"
(flow-limited) energy sources.

Internal Constraint. The maximum area A* that can be exploited given the internal constraint of power supply shortage is:

$$A * per\ capita\ =\ \pi r^2\ =\ hr/year \times \frac{C}{THT} \times \left(1 - 0.2\frac{r}{ws}\right) \times 1/\alpha \qquad (9)$$

Equation (9) is based on the rationale that only a fraction of the total human time (THT) available to a village (population $\times$ 8760 hr/year) can be actually invested in the activity of securing energy input from the environment (only the fraction "C/THT"). A further fraction of this time is lost in travel from and to the village. Such a reduction factor $[1/10 \times (10 - 2\ r/ws)]$ can be calculated based on one round trip a day and a workday of 10 hours (*ws* is expressed in km/hr, and *r* in km). The average area that can be exploited per capita depends on the requirement of actual labor hours in the field per unit of area exploited (α). Clearly, an increase in human time lost in traveling that follows an increase in area exploited per capita reduces the time available to work in direct exploitation. Hence, an increase in area exploited per capita can be achieved only by increasing the rate of power delivery during traveling or during activities of direct exploitation in the field. The more power delivered, the shorter the time required for moving around or for performing a defined agricultural task.

External Constraint. The maximum ET* that can be collected because of external constraints is:

$$ET* = \delta \pi r^{*2} = ABM \times MF \times Exo/Endo \qquad (10)$$

where πr^{*2} is the radius of the maximum area available per capita as determined by the population density (see Figure 10, part a).

Using Equations (9) and (10) as a qualitative tool to discuss the type of constraints affecting pre-industrial societies (a dynamical simulation is presented below), for marginal and semi-desert areas it is found that the lower the value of δ, the higher the value of *r*. In this situation, pastoralism can become an obligate choice to keep low α to a level compatible with the available work supply (put another way, agriculture will not provide enough return). In this situation (low δ), the constraints on labor time favor nomadism, in order to avoid loss of time in daily transport to and from the fields. Tall individuals (e.g., Tuaregs, Masai) walk faster (*ws*) than small individuals. Agriculturists, on the contrary, enjoy a higher δ, which means a smaller *r*, and hence are able to adopt more intensive techniques of exploitation per unit of area (a mix of activities implying a higher α). It is

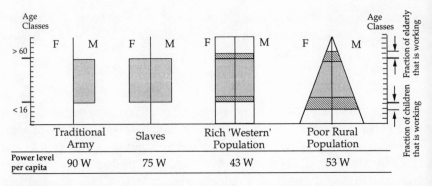

Figure 11. Population structure, sex ratio, and human power level in society.

Assumptions

Power levels:
Non-working population = 0 W; Working Population: Adult Man = 90 W; Adult Woman = 60 W;
Working elderly and youngsters = 40% reduction adult power;

Population composition:
• Traditional Army = 100% adult men, 100% working;
• Group of slaves = 100% adults, 100% working, sex ratio 1/1
• Rich "Western" Population = sex ratio 1/1, 60% working, 10% of which is elderly and youngsters
• Poor rural Population = sex ratio 1/1, 75% working, 15% of which is elderly and youngsters

advantageous for agriculturists (e.g., Yanomamö, Pigmies) to be relatively short [a low average body size ABM in Equation (10)], because this frees energy for societal activities (a higher MF and higher exo/endo ratio).

Regarding standard of living, all pre-industrial societies operate within a range of BEP that would be considered unacceptable in Western societies (a short life span, heavy labor charge in the case of high demographic pressure, child labor, poor services, and low exo/endo values). However, the internal and external constraints on the hypercycle (SEH) as defined by Equations (9) and (10) show that this is a forced situation (Giampietro et al. 1993).

In the same way, it can be shown that the age and sex structure of a population or community affect the ability to deliver muscle power per hour of activity. This is illustrated in Figure 11. A higher human power level reduces the requirement of work hours spent in ecosystem exploitation (see Figure 7; and also Giampietro et al. 1993) and, therefore, plays a role in decreasing the severity of internal constraints determined by the need for congruence between SEH and BEP. Differences in human power level can be fundamental in societies operating with a weak hypercycle in which the investment of useful energy is mainly determined by flows of human power (e.g., power bottlenecks in front of seasonal work).

Table 4. Estimated Endosomatic Energy Expenditure (EEE) of Men and Women Carrying 100 kg of Sand on a 5 m High Stairs at Different Power Levels

Number of Trips	Weight (kg) Carried Each Trip	Time to Complete Task (sec)	Gender	Estimated Endosomatic Energy Consumption*		Relative Energy Cost— Women versus Men
3	33.3	360	men	7 × BMR	198 kJ	—
3	33.3	360	women	7 × BMR	not feasible	—
10	10	600	men	4 × BMR	189 kJ	100%
10	10	600	women	5 × BMR	193 kJ	102%
50	2	1800	men	3 × BMR	425 kJ	100%
50	2	1800	women	3 × BMR	347 kJ	82%

Note: *BMR = basal metabolic rate. For men: BMR = 1.21 W/kg and body weight = 65 kg. For women: BMR = 1.17 W/kg and body weight = 55 kg (James and Schoefield 1990).

Similarly, differentiation of labor roles between genders is likely to play a role in augmenting the return of the hypercycle in pre-industrial societies. As illustrated in Table 4 and discussed in Giampietro et al. (1993), different types of work can have different energetic costs according to the gender of the worker performing the task. In order to boost the strength of the hypercycle, women should as much as possible cover repetitive activities that have a low power demand (since they consume less energy there), whereas men should cover peaks of power requirements and remain idle (possibly processing information) when not needed. Obviously, when a strong exosomatic autocatalytic loop is available, as is the case in industrialized societies, such a differentiation loses its sense. In modern societies, men and women deliver a flow of information aimed at regulating the operations of exosomatic devices. In doing that, they are perfectly equal from an energetic point of view.

Dynamic Model of Pre-industrial Societies

As noted in Giampietro (1997, this volume), when studying the relation among elements of complex systems, everything depends on everything else. Therefore, when attempting to model socioeconomic systems, it should be kept in mind that any description will be biased by the choices of the scientist made when defining the system (e.g., Why is the scientist looking at the system in the first place?). Consequently, we can not expect to have a single "correct" way to model human systems. The approach proposed here is based on the simple rationale of establishing links among different hierarchical levels of the system (different perspectives on changes in the system) by checking the congruence of events occurring on different space/time scales. In particular, we check the congruence of demand and supply of both exosomatic energy and human time in different activities performed in the socioeconomic system.

An example of a dynamic model of pre-industrial society is shown in Figure 12. The model was developed by Giampietro and Mayumi (1996) using STELLA II. It is a basic "generic" model that can be "customized" according to the particular society under study (e.g., pastoralism in semi-desert areas or shifting agriculture in a tropical forest). Parameters considered in the dynamics refer to the endosomatic autocatalytic loop of energy only (food → human labor → food), because the ability to expand the endosomatic energy metabolism with exosomatic devices is considered

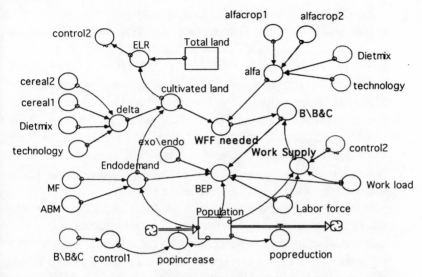

Figure 12. Examples of a dynamic model for pre-industrial society (program STELLA II).

fixed according to existing technology (see Giampietro 1997, this volume). In the model, the following parameters are defined:

1. The population size, which is expressed in number of people. The population size determines a requirement of endosomatic energy (food energy), given the average body mass (ABM) in society and the metabolic flow per kg of body mass (MF).

2. Two technological coefficients that describe the characteristics of the endosomatic autocatalytic loop. The land demand per unit of food consumed (δ), which depends on the yield per hectare for the main crops cultivated (here, we considered for the sake of clarity only two crops) and on the relative contribution of the different crops to the total food supply. The other coefficient is the labor demand per hectare (α), which depends on the labor demand per hectare for each of the crops cultivated and on the relative contribution of the different crops to the total food supply

Combined δ and α define the labor demand in food production. We can then "customize" this general model by assessing additional coefficients, such as: (a) a fixed ratio between the hours of work in post-harvest activities and the hours of work in agricultural production. This ratio indicates the number of hours that are spent in society in other activities related to food security per hour spent in direct production in the field. The product of this ratio and work demand can be used to assess total labor demand in food security; and (b) a fixed ratio between the land needed for exosomatic energy and long-term stability and the land directly exploited for food production. This ratio indicates how many hectares of land are needed for ancillary ecological processes related to food security per hectare of land actually exploited for food security in a given particular year. For example, for a 20-year cycle of shifting cultivation, the numerator would include the total area involved in the rotation and the denominator the area actually cultivated in a given year.

3. The labor supply, which is given by the following three characteristics of the social system: the fraction of total human time in society that is dedicated to work (depending on the age structure and working ages adopted in society), the work load of the working population, and the fraction of the total work time supply that is allocated to food security. Together, these three characteristics define the labor supply in food production per total endosomatic energy throughput.

4. The environmental loading, which can be assessed by the ratio between the area under direct exploitation and total accessible land (alternatively, by assessing the amount of biomass appropriated by the society expressed as a fraction of the total net primary productivity of the ecosystem over the total area supporting the society).

At this point, according to the interest of the scientist, a mechanism of control has to be defined in terms of negative and positive feedbacks. An example of the former is an increase in the demand of labor in food production when the population increases. The logic is that an increase in population results in a higher value of environmental loading EL, which in turn reduces yields and increases the labor demand per hectare. An example of a positive feedback is a positive effect of population growth on technological coefficients, resulting in higher yields and lower labor demand per hectare.

Starting assumptions are a fixed amount of natural capital (a fixed number of hectares of land with a defined net primary productivity and ecosystem type) and a set of initial values for socioeconomic parameters and technical coefficients that reflect the characteristics of the type of society considered (for more details, see Giampietro and Mayumi 1996).

Simulation of the Effects of Population Growth in Pre-industrial Society

To examine the concept of optimum population, we simulated a situation of population growth in which the population grows monotonically with a negative effect on environmental loading and a positive effect on technological performance. This simulation shows that BEP increases with the population to reach a maximum, and then steadily declines. Clearly, this is a trivial result given the assumptions used in building the model. More interesting is the ability of the model to explain several features typical of pre-industrial societies.

The ratio δ/α directly affects THT/C and, therefore, at a fixed exo/endo ratio, SEH. Hence, a maximum δ/α is the best solution to boost the energetic hypercycle. It is well known that a monotonous diet based on a cereal or a starchy root as main staple corresponds to agricultural production with a maximum δ/α.

The existence of social classes or of slaves in society can be explained as a tool for decoupling the BEP of some social groups (the ones expressing the function of control provider) and the SEH of other social groups (the ones expressing the function of operating the hypercycle). In this way, the social group of rulers can increase the work supply WS without affecting the average BEP of the society (by increasing the work load of slaves above the average and reducing their consumption below the average). Indeed, the existence of slaves and lower classes boosts the work supply per unit of energy consumed.

The concept of carrying capacity in pre-industrial society can be studied by adding a control on population that is activated when the labor demand in the primary sector is absorbing a too high fraction of the labor force (that is, when the ratio B/(B+C) becomes too small compared to a chosen "minimum acceptable standard." This is Control 1 as indicated in Figure 12. The rationale here is that a dramatic reduction in work supply allocated to nonessential activities (the saturation of work supply by the demand from CI activities) diminishes the system's buffer capacity to deal with fluctuations and perturbations that might threaten ET returns. This is discussed in detail below in the section on the problem of controlling such a dynamic system.

In this second model, a decline in fertility of the population is "activated" (through cultural control) when the labor demand in food production is absorbing too much of the available work supply (which would have the effect of reducing too much BEP). Two different threshold values have

a. Carrying Capacity when BEP* = 9500 ---> 575 people

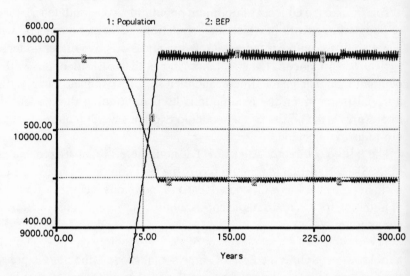

b. Carrying Capacity when BEP* = 8000 ---> 875 people

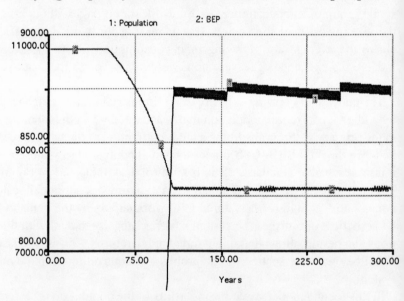

Figure 13. Example of simulation: Carrying Capacity depending
on minimum acceptable BEP.

been used to illustrate the activation of population control: (1) when food security absorbs 40 percent of the total work supply, and (2) when food security absorbs 30 percent of the total work supply.

The simulation with the first setting (40%) is shown in Figure 13, part a. The value of BEP decreases with an increase in population and then, when the fraction of the work supply absorbed by food security reaches 40 percent, cultural control stops population growth. This implies stabilizing the societal energy budget at a BEP of 9500 kcal/hour (about 40 MJ/hr) and a population size of less than 600 people.

The simulation with the lower threshold value for population control (30%) induces a stabilization of the population at a larger size—almost 900 people. This greater size is paid for with a lower value of BEP, about 8500 kcal/hour (35.5 MJ/hr), as shown in Figure 13, part b.

The two threshold values that trigger population control define two different values for the carrying capacity, 575 people in the first case and 875 people in the second. That is, carrying capacity depends on: (1) technical coefficients (the ability to achieve an equality of BEP and SEH), (2) environmental loading (the relative scale of society and the ecosystem in which the society operates), and (3) the cultural identity of society (the critical BEP or B/(B+C) value at which a negative control on population growth is activated).

Controlling Dynamic Systems: Tainter's Theory of the Collapse of Pre-industrial Societies

In the model of pre-industrial society, the socioeconomic system can not operate close to the point of equality between the C that is required and the C that is supplied [i.e., at very small values of B/(B+C)]. This would be too risky in case of fluctuations. In these conditions, the system is likely to rapidly collapse if "C required" exceeds "C supply," or to explode if "C supply" exceeds "C required." Other parameters have to quickly change to restore imbalances between requirement and supply. This explains the large fluctuations in the sizes of pre-industrial societies. Obviously, in pre-industrial times, complex large societies could not afford large fluctuations in population size and, therefore, had to constantly maintain a safety buffer. Hence, in the past, complex societies always had to assign a surplus of labor (WS − C = B) to nonessential activities, such as the construction of pyramids, great walls, and cathedrals. An excellent discussion of this point can be found in Mendelsshon (1974), which discusses the essential role of pyramid building in stabilizing the development of the Egyptian state. Cottrell (1955) calculated that the yearly endosomatic energy cost of pyramid building was approximately equal to the energetic surplus of Egyptian agriculture. Had this surplus been invested in an expansion of agricultural production, Egypt would have risked a Malthusian trap with reduced returns and excessive environmental loading that could have resulted in instability in the short term (famines and social unrest in less favorable years) and been fatal in the long term (as with the collapse of the Mesopotamian civilization because of ecological degradation).

Thus, the parameter B/(B+C) can be used as an indicator of the efficiency of the hypercycle currently adopted. In case of crisis, when an increased demand of labor for C activities reduces the ability to allocate labor to B activities, the nonessential B activities are (temporarily) suspended. The need to allocate a consistently high fraction of the available work supply to subsistence activities (C) is a signal that the system needs to undergo an adjustment that either increases SEH, reduces BEP, or alters the ratio of population/natural resources. Keeping the ratio B/(B+C) sufficiently high is a guarantee for the survival of society in the short term, for it buys time for the system of controls to adjust the parameters that determine the equilibrium between BEP and SEH in case of fluctuations. History shows that one of the most popular solutions to the problem of adjustment in pre-industrial societies was getting into war.

Tainter (1988) has studied in detail the growth and collapse of complex societies and has beautifully described how some pre-industrial societies managed to reach a certain critical level of complexity (e.g., formation of large empires), in spite of the innate weakness of their primitive exosomatic hypercycle based on fund-type energy sources, and then collapse. The weak hypercycle implied that the building of an empire had to go through an initial phase of conquering small kingdoms (exploitation of resource stocks) rather than through an increase in efficiency through scale coordination over larger areas of agricultural production. The development of very large empires was further dependent on the availability of suitable waterways to reduce the cost to the central administration of the transportation of surpluses (Cottrel 1955; Debeir et al. 1991), as well as on the exploitation of slaves and differentiation among social classes to keep high the SEH and BEP of the empire.

Regarding the mechanism of control on the stability of the dynamic energy budget, we can visualize the utilization of energy return (ET) within the socioeconomic system, as shown in Figure 14. At any moment

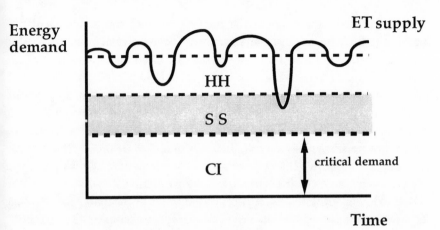

Figure 14. Socioeconomic systems as dynamical systems: The problem of control.

in time, the energetic return, ET, available to society because of its activities is distributed to meet: (1) the requirement of the hypercycle (CI), which is a "nonnegotiable" appropriation; (2) the requirement of the ruled (HH), which is the consumption of tax-paying farmers; and (3) the requirement of rulers (SS), which in complex pre-industrial societies was the consumption of army, administration, public works, and royal courts. In periods of ample surplus, the consumption of the rulers increases, thereby expanding the SS compartment, whereas in periods of shrinking surplus, the consumption of the HH sector is compressed. Tainter (1988) has provided beautiful examples of this "ratchet effect." For instance, in the Roman Empire, the ratchet effect resulted in a continuous increase in the ratio of the number of workers in the administration of the empire and the number of farmers and, consequently, in a continuously increasing tax pressure exerted on rural communities.

The typical predatory behavior over smaller kingdoms in the initial growth phase of large empires definitely enhanced the ratchet effect. Young empires were inclined to invest a sensible fraction of their energy throughput ET in an army and in administration, as investments to improve the efficiency of the army and administration paid off much better than did investments to improve the standard of living of farmers. Conquering kingdoms represented a form of resource *stock* exploitation and a means to boost the ratio of ET/CI well above that obtainable through resource *fund* exploitation (agriculture) by happy farmers. However, when the growing empire ran out of kingdoms to conquer, the burden of maintaining the expanded SS sector relied completely on the surplus generated by agricultural activities. At that moment, the gap between the requirement (BEP) of the empire and the actual return generated by its primary sectors (SEH) became evident. For a while, the gap could be filled by increasing the taxation of farmers ("squeezing" the lower-level holons) and, thus, reducing HH consumption (see Figure 14), but this solution made the daily life of farmers (lower-level holons) impossible in the long term, causing them to lose their feeling of belonging to the socioeconomic system (empire) in which they performed a vital role. In terms of hierarchy theory, the vertical coupling among levels in the holarchy was weakened by a breaking of the double asymmetry (with higher-level holons getting more than they are giving to lower-level holons). In these conditions, even small perturbations led to the collapse of the entire system. A recent example of this mechanism was the collapse of the Soviet Union after the destruction of the Berlin Wall.

Dynamic Model of Industrial Societies

Modeling Modern Developed Societies

A modern developed society characterized by a THT/C higher than 25/1, an exo/endo ratio greater than 30/1, and a BEP of over 350 MJ/hour can not be studied by the model presented in Figure 12. The existence of a well-developed autocatalytic loop of exosomatic energy in industrialized society requires that the characteristics of both the endosomatic and exosomatic autocatalytic loops be considered. We now describe a more complex model (which we simulated using STELLA II) that fulfills these criteria. We will briefly discuss its main features and then present results.

The model is divided into four compartments or sections. Section 1 includes intensive variables that define the characteristics of the exosomatic autocatalytic loop (SEH). Section 2 includes the demand and supply of both exosomatic energy and human time in the different sectors. Differences here determine the activation of controls. Section 3 includes the intensive variables that define the characteristics of the socioeconomic system (BEP). Section 4 includes extensive variables (population and natural capital) that consider the effect of changes in size of the socioeconomic system (comparing the scale of human activity with that of ecosystem activity).

The exosomatic energy compartment is defined in terms of technical coefficients that in turn determine the level of energy dissipation. Technical coefficients include: (1) the exosomatic cost of food security (FS) (ratio exo\endo of FS); (2) the exosomatic cost of environmental security (ratio exo\endo of ES); (3) the exosomatic energy spent in the service sector (based on average energy expenditure of labor in the service sector, in MJ/hour); (4) the exosomatic energy spent in household activities (based on the average energy expenditure of human time allocated to A activities, in MJ/hour); (5) the exosomatic cost of exosomatic devices (exo/exo in B&M, applied over the sum of the preceding four flows of exosomatic energy); and (6) the cost of energy and mining (exo/exo in E&M, applied over the sum of the preceding five flows of exosomatic energy). Using our database of 107 countries, we estimated: (1) range of values of the six parameters listed above; (2) expected changes in values of these parameters in relation to changes in values of BEP; and (3) a range of values of the energy throughputs per hour of labor in the four compartments of CI activities. The latter values define the aggregate demand of labor in C, given the value of the exosomatic energy spent in that sector.

Characteristics of the hypercycle (ET/DI and EETDI$_{DI}$) are determined by the quality of the energy sources and the type of exosomatic converters used (also, these values can be estimated from the database). These characteristics define the potential ET supply that is achievable by society with the labor supply in the energy sector.

The basic analysis of the dynamic energy budget was discussed in the preceding paper (Giampietro 1997, this volume). Total human time (THT) represents the potential computational capability of society (a proxy for brain ability to process and store information for decision making). The "amount" of capability can be allocated to the autocatalytic loop of exosomatic energy for the production or the consumption of goods and services and for building efficiency or adaptability. The difference with the pre-industrial case presented in Figure 12 is that, in industrial society, assessments referring to endosomatic energy (food security characteristics) are relevant only in terms of demand of exosomatic energy (investment required in CI), demand of human time, direct effect on age structure and morbidity (health effects of nutrition), and direct effect on the environment (environmental effects of agriculture). Hence, in industrial society, endosomatic energy is *not* considered as constituting an autocatalytic loop.

Labor supply is given by the characteristics of the socioeconomic system. Total work supply (B+C) and work supply for primary sectors (C) are defined according to the following parameters of the model: (1) fraction of the population that is economically active; (2) work load (yearly labor charge for labor force); (3) unemployment rate (expressed as a fraction of the working force); and (4) the defined desired ratio B/(B+C), which is a given minimum acceptable value determined by the cultural identity of the society.

The fourth sector of the model, the only one with extensive variables, is used to deal with the effects of changes in scale. These include positive effects related to economies of scale and specialization, and negative consequences for SEH because of excessive environmental loading. The scale of socioeconomic activity (ET) is given by the product of three terms: population size, MF×ABM, and the exo/endo ratio. Environmental loading is assessed by a ratio of ET and an indicator of environmental size. The latter indicator should be a proxy of the scale of biophysical activity providing life support to the society. Hence, the ratio can be simple, such as ET/area in J/year/ha, or sophisticated, depending on the indicator of biophysical activity chosen.

The choice of a mechanism of control regulating the industrialized socioeconomic system (that is, the definition of which parameters are affecting what other parameters, on what time scale, and by which measure) depends on the aim and interest of the scientist performing the analysis. This choice involves an unavoidable element of arbitrariness but, of course, the credibility of the assumptions regarding the mechanism of control can easily be checked by other scientists even when they work in different fields or have different interests. In fact, the numerical values resulting from the simulation have to be consistent with scientific findings about the same parameters in other disciplines which study socioeconomic systems from different perspectives.

Regarding the mechanism of control, we considered two negative feedbacks and one positive feedback in the model. The negative feedbacks involve an automatic negative effect of scale on the value of ET/CI (an increase in population size results in an increased size of ET, which in turn augments the environmental loading, which in turn induces a reduction in ET/CI). Negative feedbacks also include a cultural control on fertility operating at a defined threshold (e.g., when the work demand in the primary sectors of the economy absorbs more than 50 percent of the available work supply, society somehow reacts by halting population growth). The positive feedback concerns an increase in the size of population, or ET, that is defined to translate into a higher value of CI/C.

Two simulations using this model are presented in Figures 15 and 16. From these data, we developed the following considerations. First, a system of control based only on a negative feedback on population control (with the contrasting effects of scale) is too coarse for this model. The autocatalytic loop "value of CI affecting ET and value of ET affecting CI" makes the behavior of the system extremely sensitive to changes in initiating conditions. Small changes in initial population size or in the shape of the curve CI/C (ET) can bring the system into areas of chaotic behavior. When the simulation is extended over a longer period, the system tends to oscillate around possible solutions (stabilization of population size around a defined level of BEP) and then enters into a chaotic behavior and finally collapses.

In fact, several other biophysical controls can be imagined to better stabilize the model by using one or more of the following parameters: (1) the work load of the labor force (this can be changed with a quite short lag time); (2) the ratio of working/non-working population (it changes automatically with a lag time of 15-20 years when the rate of population growth changes, and can also be changed with a constant population structure by altering the age limits for entering and leaving the labor force); (3) level of education (this can be manipulated by defining different age ranges for compulsory education and controlling access to higher education); and (4) unemployment rate (unemployment can play the same role of buffer and signal that we chose for the ratio B/(B+C) in the model of pre-industrial society).

A second consideration is that, because of trade, it is difficult to define the size of the natural capital on which a defined industrial system is dependent. Trade can make available a large amount of natural capital embodied in imported raw materials, commodities, and services. Therefore, a society with a strong economy can escape environmental constraints by transferring its environmental loading to other (often less developed) societies.

Last, in modern societies it is the economic process that regulates the system in the short term. In fact, economic processes react so quickly to changes induced by biophysical constraints that often the biophysical reading of the economic process is obscured. For example, in a developed society based on trade, ET/CI depends on prices and added value generated in the primary sectors rather than on actual technical coefficients. Moreover, incurring debt (e.g., a growing national deficit) can boost ET/CI and cause the parameters B/(B+C) and BEP to reach effective values that are higher

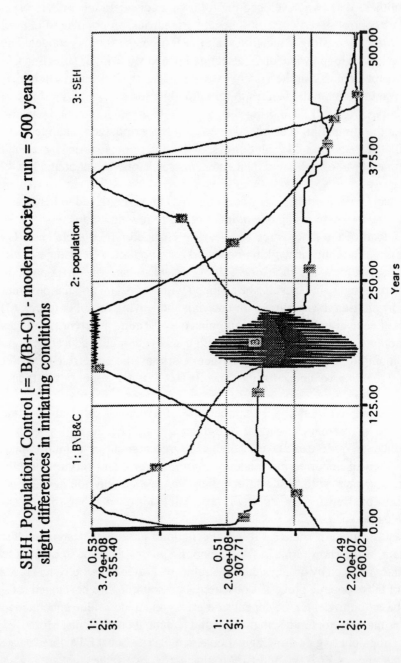

SEH. Population, Control [= B/(B+C)] - modern society - run = 500 years slight differences in initiating conditions

1: B\B&C   2: population   3: SEH

Figure 15. Simulation (using STELLA II) of dynamic evolution of a developed society.

170

SEH. Population, Control [= B/(B+C)] - modern society - run = 1000 years slight differences in initiating conditions

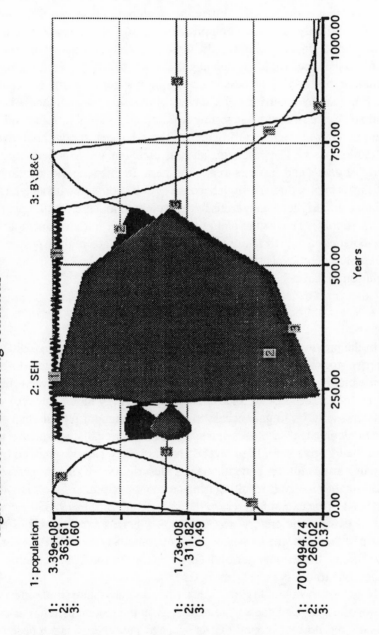

1: population   2: SEH   3: B\B&C

1: 2: 3:	3.39e+08 363.61 0.60	
1: 2: 3:	1.73e+08 311.82 0.49	
1: 2: 3:	7010494.74 260.02 0.37	

0.00   250.00   500.00   750.00   1000.00

Years

Figure 16. Simulation (using STELLA II) of the appearance of chaotic behavior.

than the "limits" imposed by SEH. This is particularly important in modern societies, in which any increase in the price of fossil energy is reflected by a net reduction of SEH. Where the discrepancy between SEH and BEP would normally result in a compression of B/(B+C) below the acceptable minimum defined by cultural identity and social aspirations, debt can prevent such a "recession" by keeping the ratio of ET/CI artificially high (for example, by enabling access to energy inputs from the international market). In fact, by incurring debt, society obtains access to oil and other exosomatic devices without investing a fraction of its ET to mine and refine the oil, without manufacturing the required devices, and without investing a fraction of its ET in producing an equivalent amount of added value in the form of goods and services to pay for them. In this scenario, a ratio of ET/CI that is kept artificially high because of debt, coupled with a high ratio of CI/C, could trigger a dangerous hypercycle (too much ET return per unit of work supply). This can explain the resurgence of massive unemployment in societies in which national deficits and debt are high. An increase in unemployment levels (fast reduction of C/WS) can be seen as an obligate solution to dampen such a hypercycle.

The "Faustian" Solution: Sustaining BEP in Developed Societies

In the previous section, we saw that trade coupled to the exploitation of energy stocks rather than energy funds allows modern society to bypass some biophysical constraints regulating the equilibrium between BEP and SEH. In fact, because of trade and stocks of energy inputs, modern societies can use IOUs to gain access to energy throughput ET. In other words, when biophysical surpluses of energy are economically unavailable to sustain the FI sector or a high level of welfare, they can be obtained by borrowing from future generations. A description of the hypercycle of exosomatic energy as resulting from an economic point of view is shown in Figure 17 (the example refers to the U.S. economy). In 1988, the cost of 1 GJ of exosomatic energy was U.S.$4.8, whereas for each GJ of energy spent in that year, the U.S. economy generated $57.7 of GNP. Under these conditions, the throughput of 1 GJ of energy in the U.S. economy made accessible to the system 12 more GJ.

Note, however, that Figure 17 describes a steady-state situation in which flows of money and energy are regulated by the mechanism of the market. Hence, the flow of money that is spent to buy energy is supposed to be actually produced in terms of goods and services sold. This is not an accu-

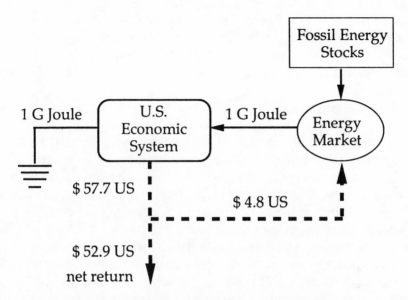

Figure 17. Hypercycle of fossil energy (stock-type energy sources) assessed from an economic perspective.

rate picture of modern economies that run on debt. An increase in monetary debt has an amplified effect on the expansion of economic activities and on the standard of living: It drags more oil into the economic process. For example, the yearly debt per U.S. citizen increased from $8,821 in 1984 to $10,036 in 1985 (U.S. Bureau of the Census 1990). This increase is equivalent to about 8 percent of the GNP per capita, or nearly the amount of money that was required per capita to pay the energy bill of that year (see Figure 17). No wonder that neoclassical economics does not account for energy among the production factors. The quality of fossil fuels as an energy source is so high that a small change in a country's debt structure can provide its energy consumption for an entire year for free (for the present generation). If this mechanism is used as a general solution in developed and developing countries—as seems to be the case—to sustain or improve the ratio ET/CI and, therefore, the share of useful energy that can be allocated to the FI sector, it is probable that eventually nobody will pay back these debts. In fact, the more that the better energy stocks are depleted first, the greater will be the diminishing marginal returns of the energy supply system (ET/CI) in biophysical terms (Hall et al. 1986; Gever et al. 1991). This implies that less useful energy surplus will be available to pay back with extra goods and services the IOUs used to obtain the energy consumed—unless a dramatic setback in the standard of living causes a sharp increase in the prices of future goods and services. Therefore, the present generation is shifting to the future not only the environmental cost but also the economic cost of overexploiting nonrenewable energy resources.

In conclusion, Western civilization has undergone an "ecological release": Through stock depletion, ecosystem degradation, reliance on imports from distant ecosystems (transferring environmental loading to other societies), and debt making, modern societies have been able to adjust the equilibrium between BEP and SEH without paying much attention to the existence of external constraints (critical thresholds of environmental loading) or internal constraints (biophysical characteristics of the autocatalytic loop of exosomatic energy). Framing this issue in ethical terms, we may say that the high material standard of living achieved today by some social groups of the present generation is based on an uneven distribution of returns within countries and among countries and on borrowing from future generations by an overdrafting of ecological services.

TUTORING THE ENVISIONING OF FUTURE SCENARIOS

A Need for Envisioning

The second half of this century has seen an impressive acceleration of changes in technological coefficients and cultural identity for many socioeconomic systems, notably in the Western world. These rapid changes have been accompanied by a blurring of the relation between socioeconomic characteristics and technical coefficients on the one hand and the availability of natural resources on the other hand. This relation, once so clear, has become hidden by the buffering mechanisms on which modern civilization relies, such as stock depletion, international trade, national debts, abuse of the resilience of ecosystems, and uneven distribution of returns of useful energy. These profound changes in societal organization have been occurring on a time scale of no more than a few decades. This is a time period too short for the systems of control in society to test whether or not the information, strategies, and positive and negative feedbacks currently used to regulate societal activities are effective in directing technological development according to our aspirations and ecological compatibility.

The last decades have made it evident that modern societies are experiencing serious problems with the management of their technological development and that there is a need to find new and different paths of development able to achieve a better compatibility with ecological systems and human aspirations. This situation has stimulated the exploration of new methods for dealing with technological development. One of these, which has rapidly gained popularity in the last years, is called "envisioning of our common future" (e.g., Meadows 1994). Envisioning means starting with the picture of the society we would like to live in and then designing technologies and development strategies for getting there.

To be sustainable, the technological development of society should fulfill two basic conditions. First, it has to be compatible with external boundary conditions. In biophysical terms, this means that the requirement of resource inputs and the generation of wastes by society must be compatible with the ecological processes that provide the resources and life-support for the people in society. In economic terms, it means that society must perform a set of activities that results in economically viable interactions with other socioeconomic systems. Second, it has to be compatible with the cultural identity of the people composing the socioeconomic system. That is to say, socioeconomic characteristics, such as material standard of living and distribution of wealth, as well as political characteristics, such as power distribution, must be compatible with the aspirations, values, and mores of the people living in society. Given these basic conditions, the idea of using envisioning to guide technological development according to the future states we want to reach seems to make much more sense than the current solution of looking for technological fixes to old problems and then passively waiting to discover what consequences these technological fixes will induce on the future quality of our life.

In general, as discussed in the preceding paper (Giampietro 1997, this volume), modeling societal development is anything but easy, and it is therefore important to recognize our limitations in predicting future scenarios. It follows from the complex nature of real-world processes that we inevitably have to face uncertainty in the discussion about future scenarios. Moreover, human societies have a special status within the class of complex systems: They are self-aware systems and, therefore, characterized by the continuous generation of emergent properties and true novelties (Funtowicz and Ravetz 1994). In the debate on the sustainability of human development, this uncertainty has been heavily employed by "technological optimists" (also called Cornucopians) to reject pessimistic views of our common future formulated by "neo-Malthusian pessimists" (e.g., Simon 1981; Abernethy 1991). Indeed, at the heart of the neoclassical economic paradigm is a blind faith that human creativity will be able to substitute limited resources and to maintain the status quo in spite of changing boundary conditions. In this context, the call for "envisioning" is an attempt to use the power of human creativity in a different way. With envisioning, this power is used to choose paths of development that we really want rather than to desperately look for quick fixes to get out of trouble when facing the bottom of the barrel.

Envisioning can be used as a tool to fight "cultural mode-locking" that prevents us from choosing more eco-compatible and human-friendly patterns of interaction between society and environment, toward which society might evolve. In fact, envisioning implies that a discussion about sustainable development should start by defining what we, as a social reality, would like to become in the future, by envisioning our daily life, the future daily life of our children, and the future characteristics of the environment we would like to live in and of the society that would match our aspirations as human beings. The rationale of this approach is that the discussion on how to implement sustainable development should start by choosing a "virtual future" to pull present decisions, rather than by allowing the "given past" to push present decisions (Weber 1996). Envisioning correctly focuses on the crucial role of the participation of the stakeholders (the people and different social groups involved in the process of development) in the entire process, including the formulation of the terms of reference for discussing future changes.

In this context, there is another dimension of the power of human creativity to be considered, for which we propose the term "passional entailment." "Passional" indicates anything affecting human behavior that is not "rational," whereas "entailment" makes reference to the Rosen's concept about the mechanism stabilizing resonating systems, as discussed in Giampietro (1997, this volume). Passional entailment represents the ability of human systems to generate a "virtual future" based on the common aspirations of the people composing a particular socioeconomic system. Such a "virtual future" can actually affect the evolution of the system from its present state (this is also referred to as self-fulfilling prophecy). Given the fact that the status quo can not be maintained forever (and that many in a given society do not like to maintain it), we can use the power of human creativity to "envision" different paths of development (Meadows 1994).

The model presented in the preceding paper (Giampietro 1997, this volume) can contribute to a better "envisioning" by performing basic reality checks on future scenarios formulated by "passional entailment" (by checking whether our aspirations clash against basic conditions of sustainability). The goal here is to reach a better mix between passion and rationality when discussing possible future scenarios.

The Model as a Tool to Perform Reality Checks on Future Scenarios

Recalling the scheme of analysis presented in Giampietro (1997, this volume), we observe that any particular configuration of society is subject to a set of basic biophysical constraints regarding the profile of allocation of energy and human time to socioeconomic activities. These internal constraints can be quantified and used to check the feasibility of hypothetical configurations of the socioeconomic system. In particular, the primary sectors of the economy require a certain minimum fraction of the energy throughput in society for their operation. The primary sectors depend on the efficiency of the food system, the efficiency in building exosomatic devices, the quality of energy and mineral resources, and the energy cost of guaranteeing environmental security. Further, for the primary sectors to function requires a minimum fraction of total human time that depends on the four efficiencies listed above as well as on the exosomatic energy throughputs in these four activities.

These constraints on the allocation of useful energy and human time are related to the level of technological development of society and the endowment of natural resources. For instance, if it is desired to raise the material standard of living (bioeconomic pressure BEP) of a subsistence society, one should first compare the characteristics of the desired hypercycle of exosomatic energy (needed to achieve equilibrium between SEH and BEP) with the available technology and natural resources. The transition from subsistence to developed society is generally characterized by a dramatic increase in energy throughput ET (exo/endo × MF × population mass) and in investment in adaptability, in terms of both energy and human time (higher ratios FI/ET and THT/C). A greater investment in adaptability is possible only when the efficiency of the exosomatic autocatalytic loop can be improved (higher ET/CI) and an adequate amount of natural resources can be made accessible to society. If this is not the case, aspirations will clash with harsh reality, and noble intentions will cause more harm than good to the subsistence society. Hence, prior to bringing a subsistence society outside its basin of stability (where a low BEP is equal to SEH and compatible with ecological constraints), it is important to check whether the proposed scenario is consistent with thermodynamic constraints.

It should always be kept in mind that a dramatic increase in BEP is linked to the ability to decouple the profile of energy allocation from the profile of time allocation among the various sectors of the economy (primary sector, service sector, and households). A large difference in these profiles implies the existence of strong biophysical constraints: Sectors of the economy that have a fraction of energy allocation that is much higher than the fraction of human time allocation must necessarily have a high labor productivity. Notably, a large dependency ratio, a high per capita resource consumption, and a well-developed service sector are all typical of a society with a high BEP. Together, they imply the availability of technologies and natural resources that enable huge throughputs of both energy and matter flows per unit of labor time in the primary economic sectors. Two examples are presented below.

Reality Check on Alternative Energy Sources

Internal Constraint: Labor Productivity in the Energy Sector

The energy sector of society must be able not only to make accessible to society an adequate flow of energy input (e.g., being able to extract or import sufficient fossil energy) but also to transform the energy input into useful energy and then distribute it while using only a small fraction of the available labor supply. This implies the existence of an internal biophysical constraint (imposed by BEP) on labor productivity.

The minimum exosomatic energy throughput in the energy sector (EET^*_{EXO}) for a society operating at a defined BEP can be expressed as:

$$EET^*_{EXO} = BEP \times (WS/C_{EXO}) \times (C/WS) \tag{11}$$

where WS/C_{EXO} is defined as the inverse of the fraction of the labor force employed in the energy sector of society, and C/WS is the fraction of the labor force employed in the primary sectors defined as CI activities.

In general, the present Western standard of living (BEP) is based on a throughput higher than 400 MJ of commercial energy consumed by society per hour of labor in the primary sectors of the economy. Since the labor engaged in the energy sector is generally less than 2 percent of the total work force ($WS/C_{EXO} = 50$) and the primary economic sectors absorb less than 50 percent of the work force ($C/WS = 0.5$), it is evident that in order to achieve such a result the throughput per hour of labor in the energy sector must be in the order of 10,000 MJ/hour of labor.

For example, in Italy, with a population of 57 million, only 7.3 percent of the total of 499 billion hours of human time available in a year were spent in paid work in 1991. Of this yearly labor supply, 60 percent was absorbed by the service sector, 30 percent by the industrial sector, and nine percent by agriculture, fishery, and forestry, leaving a tiny one percent, or 360 million labor hours, to run the entire energy sector (ISTAT 1992). Total energy consumption in Italy that year was 6,500,000 TJ, implying that in 1991, the Italian energy sector delivered almost 18,000 MJ of energy throughput per hour of labor in that sector. This throughput was achieved while using mainly (about 90%) fossil energy sources.

Energy resources with a much smaller energy throughput per hour of labor (e.g., biofuel such as ethanol, with a maximum of 1,600 MJ/hour) would require that between 20 percent and 40 percent of the labor force of society (after absorbing all the unemployed) be allocated in the energy sector, which is incompatible with the current profile of labor allocation to the various economic sectors (Giampietro et al. in press). This constraint is totally overlooked in most discussions of alternative energy technologies, especially energy from biomass. Envisioned future societies cannot rely on biofuel if they want to operate at a BEP over 300 MJ/hour (can not have a large service sector, people with a long life span, and so forth). The same applies to energy sources based on the recovery of wastes and by-products (e.g., biogas). When the energetic remunerativity of recycling—that is, the energy gain obtained by recycling divided by the extra labor required—is lower than the minimum labor requirement per net GJ imposed by socioeconomic constraints, recycling should be considered a service and, therefore, a cost.

External Constraint: *Ecological Compatibility of the Energy Sector*

When ET is generated from a fund-type rather than a stock-type energy resource, the total annual energy demand implies the requirement of a certain amount of land, fresh water, and other environmental services. This "space constraint" can be formulated as follows:

$$ET = \delta \times exploited\ area \times ecological\ buffer \qquad (12)$$

Where:

δ is the density of the energy input made available by natural processes and harvested per unit of area and per unit of time (e.g., measured in W/m^2);

Exploited area is the area directly managed by humans by means of application of useful energy (where the economic process results in direct alteration of natural patterns of biophysical flows of energy and matter);

Ecological buffer is a multiplication factor accounting for the area that should be left in wild configuration to guarantee the stability of ecological communities and the life support system of society. It is calculated as (wild area + exploited area)/exploited area. This ratio is typically about 20 in pre-industrial agricultural societies of swiddeners. E.P. Odum (1971) suggests that 1.33 is the minimum value of ecological buffer that would be required worldwide for preserving biological diversity. Within the Odum school, Brown et al. (1992) propose that the buffer ratio should depend on the environmental loading ratios generated by society and by the fragility of the ecosystem altered. According to the latter approach, the ecological buffer increases with the intensity of human interference on biophysical processes within the altered area and the fragility of the ecosystem in which society is operating.

Energy resources operating with a requirement for environmental services (e.g., land and fresh water demand) incompatible with the current boundary conditions are clearly not feasible. For example, biofuels with a requirement of hundreds of tons of fresh water and between 0.04 and 0.14 hectares of arable land per net GJ of fuel delivered to society are unlikely to represent a viable alternative to oil, on a large scale, in the future (see Giampietro et al. in press).

Reality Check on Providing Food Security

Internal Constraint: Labor Productivity in the Food System

Biophysical constraints also exist for achieving food security in society. Again, a society must not only have access to an adequate flow of food input (through import or production) but also be able to make it accessible to the final consumer (processing, packaging, distributing) by using only a small part of its labor force. An equation of balance can be formulated either for the food system as a whole or for the agricultural sector only. In a country that desires to be self-sufficient with respect to food production, the constraint related to endosomatic energy—that is, the minimum endosomatic energy throughput in the agricultural sector (EET^*_{ENDO}) can be written as follows:

$$EET^*_{ENDO} = (MF \times ABM) \times P/C \times THT/WS \times WS_{AG}/WS \qquad (13)$$

In Equation (13), P/C is the ratio between agricultural production produced and consumed. It measures the fraction of primary agricultural production needed in excess of "MF × ABM" (endosomatic energy requirement per capita) to compensate for post-harvest losses and double conversions (e.g., production of grains for meat production). The term "MF × ABM × P/C" defines the required energetic output of food in the agricultural sector per capita. THT/WS is the inverse of the fraction of total human time spent working. WS_{AG}/WS is the fraction of the labor force engaged in the agricultural sector of the economy.

For example, in 1994 the United States had a population of 260 million people corresponding to 2,277 billion hours of total human time. However, only 49 percent of this time (1116 billion hours in 1994) referred to the economically active population. Due to sleeping and leisure time, only 20 percent of these hours (223 billion hours in 1994) were spent in actual work. Of these, only 2 percent were allocated to the agricultural sector (4.5 billion hours). Therefore, the amount of food consumed per U.S. citizen in 1994 was generated by about 17 hours of work in the agricultural sector during the entire year. Thus, to maintain the actual distribution of the work force among different economic sectors, a developed economy, such as that of the United States, must have the technology to produce, on average, a minimum of 490 MJ of food per hour of labor (assuming P/C = 2, which is likely to be an underestimation). Such a labor productivity is well out of the reach of farmers in all developing countries (e.g., farm labor productivity is less than 4 MJ/h in China) and is not even reached by farmers in the European Union (whose productivity is less than 100 MJ/h) (Giampietro in press). The fraction WS_{AG}/WS is about 0.02 in the United States and Canada, and generally is smaller than 0.07 in developed countries. Techniques of agricultural production that do not use mechanical devices and heavy energy subsidies (such as the ones adopted in subsistence societies) have a much lower productivity of labor and, therefore, imply a huge demand for labor in the agricultural sector. No developed country could operate with 30-40 percent of its labor force engaged in the agricultural sector. This constraint deserves attention in discussions of alternative technologies for the agricultural sector, such as low-input agriculture, mixed cropping, and gardening.

External Constraint: Space Requirement for Food and Environmental Security

Industrial societies based on fossil energy have a certain margin for dealing with land constraints for their food security. In fact, they can boost land productivity by relying on the heavy use of inputs or can import food commodities from elsewhere. However, a biophysical constraint referring to space remains because of parallel requirements of space for food production, housing, and infrastructures; for matter cycling (extracting raw materials and dumping wastes); and for guaranteeing environmental security.

Constraints referring to the requirement of space for food production are certainly crucial, inasmuch as more than 99 percent of the food consumed worldwide is from land. The constraints have been analyzed in detail elsewhere (e.g., Kendall and Pimentel 1994; Ehrlich et al. 1993). Basically, an analysis of space constraints can be carried out by using equations similar to Equation (12).

The space constraint related to environmental security can then be formulated as:

$$EL_{ES} = (ET/exploited\ area) \times (1/ecological\ buffer) \le EL_{max} \qquad (14)$$

The ecological buffer is determined by the area that must be set aside from high-intensity energy use and in which natural energy flows should be left as close as possible to natural densities in order to keep the environmental loading (EL) within a range of values compatible with the stability of the ecosystem in which the society is operating.

When the indicator EL_{ES} exceeds a critical threshold, the scale of the economic process has become too large compared to that of the biophysical processes sustaining it. This threshold can be detected by the appearance of two negative consequences. First, on the side of human society, a too-large size of the economic process increases the exosomatic energy expenditure for making accessible the same flow of matter throughput (e.g., 1 kg of corn, 1 liter of clean water). That is, a further expansion of the economic scale generates a decreasing return of the energetic investment (it reduces the fraction of disposable income that can be maintained per unit of ET). The decreasing returns may be observed as: (1) a decreasing efficiency of the food system, (2) a lowering of the quality of energy and mineral resources, and (3) the addition of a new set of activities, such as pollution control and environment monitoring and restoration, to the set of CI activities. The threshold is detected by a decrease in the ratio of ET/CI in terms of allocation of useful energy investment, and a decrease in the ratio of THT/C in terms of human time investment.

Second, on the ecological side, a too large economic process decreases the stability of natural systems by decreasing the capability of solar energy to maintain the structure and functions of the ecosystem. Put another way, a reduction in the autocatalytic loop of energy which is transformed under natural biophysical control reduces the supply of ecosystem services, per unit of space and time, available to society. An excessive demographic pressure (the need to achieve high yield per hectare), especially when coupled with a high bioeconomic pressure (the need to achieve a high productivity of labor in food production), implies a high environmental impact and a high energetic cost of agriculture.

In conclusion, the goal of feeding 10 or 12 billions of wealthy people in the year 2050 translates into the need to produce sufficient food using about 0.1 hectares of arable land per capita and less than 5 percent of the work force in agriculture. A technological solution for this challenge, if possible at all, is very likely to clash with the need to preserve the biodiversity and stability of natural ecosystems. If, in the 1950s, policymakers and scientists had only organized a few sessions of "tutored envisioning" about food production for humankind in the 1990s, they probably would not have lent support to the Green Revolution in the way it was actually carried out.

DRIVE OF MODERN CIVILIZATION TOWARD UNSUSTAINABILITY

Jevons' Principle and the Myth of the Dematerialization of Economies

The last application of the model concerns the existence of a built-in mechanism that drives evolving systems toward unsustainability. This mechanism is closely related to the need for complex systems to continuously adapt; it was discussed in the first two applications in this paper regarding societal development. To adapt, complex evolving systems are forced to continuously become something else, and this provides an explanation for the counterintuitive behavior of evolving systems—such as Jevons' paradox.

Jevons' paradox (Jevons 1990) was first enunciated by William Stanley Jevons in his book *The Coal Question* in 1865. Briefly, it states that an increase in efficiency in using a resource leads to an increase in the use of that resource rather than a reduction in its use. At that time, Jevons was discussing the trend of future coal consumption. Also, at that time, others were predicting a future reduction in fossil energy consumption because of technological progress (such as more efficient engines).

Jevons' principle proved to be true not only with regard to the demand for coal and other fossil energy resources but also with regard to the demand for food resources. Doubling the efficiency of food production per hectare over the past 50 years (due to the Green Revolution) has not solved the problem of hunger. It has actually made it worse, as hunger increased proportionately more than the population size (Giampietro 1994a). In the same way, doubling the area of roads did not solve the problem of traffic; it made it worse, because it encouraged the use of personal vehicles (Newman 1991). When more-energy-efficient automobiles were developed as a consequence of rising oil prices, American car owners increased their leisure driving (Cherfas 1991). Not only did the number of miles increase, but the expected performance of cars also grew: An increasing number of U.S. citizens are now driving minivans, pick-up trucks, and four-wheel drives. In the same way, more efficient refrigerators have become bigger (Khazzoom 1987). In economic terms, we can describe these processes as "increases in supply that boost the demand."

Technological improvements in the efficiency of a process (e.g., increase in miles traveled per gallon of gasoline) represent improvements in *intensive variables*. As soon as such technological improvements are achieved, room is generated for an expansion of the size of the system (e.g., more people make more use of their cars). Lateral expansion represents a change in *extensive variables*—that is, the dimension of the process.

Unless a comprehensive analysis of technological improvement is performed, which means including intensive and extensive variables and the contrasting perspectives of higher- and lower-level holons, it is easy to be misled by the counterintuitive behavior of evolving complex systems. As an example, we examine the effect of the oil crisis of the seventies on the economy of developed countries. The immediate effect of the oil crisis on Western economies was a decoupling of energy consumption in the economy with the GNP that was generated. The economies of developed countries quickly adjusted to higher oil prices by increasing the efficiency of the conversion of energy inputs into its various end-uses (useful energy). This was obtained mainly by altering the mix of fuels used (more natural gas and oil and less coal) and improving the conversion process of energy input into useful energy. In other words, short-term technological fixes were able to generate in industrialized societies similar amounts of useful energy (maintaining the same set of activities) while consuming less energy input than before (Gever et al. 1991; Kaufmann 1992). However, when we assess the result of such a process in the long term, we find that instead of inducing a decline in energy consumption in developed economies such as the United States, this solution simply increased the demand for high quality energy sources (natural gas and oil) and, as a consequence, implied a stronger dependence on the international market for oil supplies.

This process can be analyzed for the United States using the two graphs provided in Figure 18. From the upper portion of the graph, it is clear that the decade following the oil crisis indeed showed a decoupling of the energy consumed in the U.S. economy and economic performance (GNP per capita). However, the economic definition of energy efficiency (MJ of energy consumed per unit of GNP), based only on the use of intensive variables, can be misleading. It can generate the false impression that technological progress reduces the dependence of modern economies on exosomatic energy and that, therefore, technological progress is a valid tool to deal with future problems of energy shortages. If it is true that the economic energy efficiency (assessed by the ratio of energy consumed in the economy per dollar of added value generated by the economy) increased after the oil crisis, it is also true that this has no relevance in the determination of the aggregate energy consumption of the United States (an extensive variable). In 1991, the United States operated at a much lower ratio of energy consumption per unit of GNP produced than, for instance, the People's Republic of China (12.03 MJ/USD versus 69.82 MJ/USD, respectively), *but because of that*, the United States managed to have a GNP per capita much higher than China's (22,356 versus 364 USD/year, respectively) (World Resources Institute 1994). When the energy consumption per unit of GNP (economic energy efficiency, an intensive variable) is multiplied by the GNP per capita (an extensive variable), one finds that in spite of the significantly higher economic energy efficiency, the energy consumed per U.S. citizen is 11 times higher than that consumed by each Chinese citizen.

In complex societies, such as the United States, the low ratio of energy consumption per unit of GNP reached depends on the high GNP per capita achieved, and vice versa. In these societies, the FI activities (work in the service sector and consumption) are the ones generating the larger part of the added value. The upper graph in Figure 18 clearly shows the inverse relation between energy consumption per unit of GNP and GNP per capita.

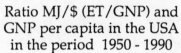

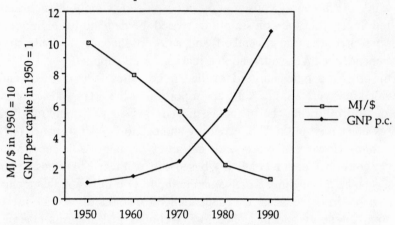

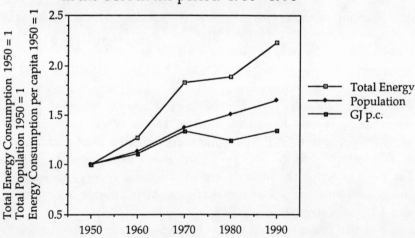

Figure 18. Intensive and extensive variables describing changes in GNP, GNP per capita, energy consumption, and energy consumption per capita in the United States during the period 1950-1990 (normalized values).

On the other hand, the lower graph shows that the temporary reduction in energy consumption per capita obtained after the oil shock in the 1970s (due to improved efficiency in primary sectors) was transformed, after a decade, into a larger investment in the service and household sectors. This internal change in the pattern of energy use (a stronger activity of the FI sector) can be detected by the dramatic increase in added value per unit of energy consumed by the system.[2] It shows that human labor allocated to the service sector generates more added value per unit of energy throughput than human labor allocated to production in the primary sectors of the economy. However, a society can shift labor time to the service sector and expand the set of activities during leisure time to increase the generation of added value only when the primary needs of its citizens have been secured. To achieve the latter with little working time requires a strong energy investment in the primary sectors of the economy. Our model indicates that a society with a high economic energy efficiency (low energy consumption per unit of GNP) must, therefore, operate at a high GNP per capita. It must already have strong efficiency in the primary sectors (very high CI/C) and, therefore, it is more, not less, dependent on huge flows of exosomatic energy input (high exo/endo) than are societies with lower levels of technological development.

The degree of "dematerialization" assumed to be induced by technological progress in the U.S. economy can be checked by analyzing the data of aggregate energy consumption published by the U.S. Bureau of the Census (1991). A reduction in the energy consumption per unit of GNP from 113 MJ/USD in 1950 to 25 MJ/USD in 1990 (a decrease of 450%) had the effect of *increasing* the aggregate consumption of exosomatic energy in the U.S. economy from 34.5 TJ in 1950 to 77.0 TJ in 1990 (an increase of 100%). As indicated by the bottom graph in Figure 18, the aggregate energy consumption increases not only because of an increase in consumption per capita but also because of an increase in population size. The latter phenomenon is explained also by immigration, driven by an attractive economy, which has the potential to counteract any decrease in the fertility rate that may occur. In fact, strong gradients in standard of living (BEP) among countries—generated by gradients in efficiency—drive labor from poor to rich countries (Giampietro and Bukkens 1996). For example, the dramatic improvement in energy efficiency that the state of California has achieved in the past decade (in terms of the intensive variable useful energy/energy input) will not necessarily curb total energy consumption in that state. In fact, present and future technological improvements are likely to be nullified by the dramatic increase in immigration, both from outside and inside the United States, which makes the Californian population among the fastest growing in the world.

Jevons' Principle and the Evolution of Complex Systems

Using hierarchy theory and the framework of analysis provided in this paper, Jevons' principle can be explained as follows: Improved efficiency of the technological processes that sustain society (e.g., more efficient light bulbs) can be transferred two ways. One is to individual human beings (the lower level) in the form of a better standard of living (BEP), while maintaining the initial level of natural resource consumption at the societal level (environmental loading EL). A lighter electric bill will enable a family to use the saved money for something else, so that the energy saved on the light bulb is spent in other activities. A second transfer is to the ecosystem (the higher level) in the form of a decrease in the consumption of natural resources by society as a whole (EL), while maintaining the initial standard of living for individuals (BEP). To obtain this, the lighter electric bill should be linked to a tax proportional to the saving, in order to prevent the consumer from using the surplus money to fuel other activities. These two possible outcomes of technological improvement are described in Figure 19, part a, in the plane BEP-EL, assuming point 1 as the initial position of the system. Following technological improvement that increases its efficiency, the system can move from point 1 to point 1' when the autocatalytic loop of exosomatic energy provides better services to humans (better BEP), while keeping the demand on the environment constant (same EL); or from point 1 to point 1" when the autocatalytic loop of exosomatic energy provides the same level of services for humans (same BEP) with lower demand from the environment (lower EL). Since human beings generally have more weight in political decision making than the environment, society tends to take advantage of technological improvements by improving the welfare of its members rather than reducing the pressure on the environment (Giampietro 1994b). As noted before, the only way to use a technical improvement in efficiency to get from point 1 to point 1" is to tax energy-efficient appliances and use such revenue only for expanding natural reserves (to prevent more human development in the long term). Clearly, the majority of voters would find it difficult to accept such a solution.

Up to now, technical progress has followed the path $1 \rightarrow 1' \rightarrow 2 \rightarrow 2' \rightarrow 3$ with alternate increases in intensive and extensive variables driven by

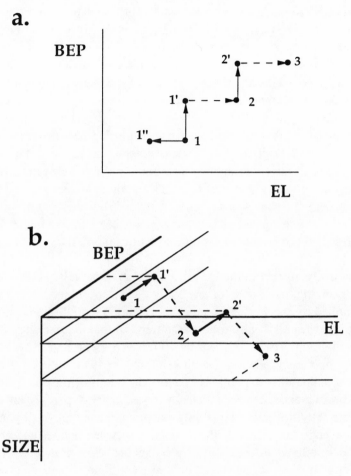

Figure 19. Jevon's principle described on the plane BEP-EL
and in the space BEP-EL-SIZE.

internal pressure (that is, with gains transferred to humans rather than to natural systems). This means that, in their evolution, human societies move upward and to the right in the plane BEP-EL. When extensive variables are also included in the picture, evolution tends to follow the trajectory 1-2-3 represented in the three-dimensional space BEP-EL-SIZE illustrated in Figure 19, part b. This trajectory moves society toward higher values of BEP and EL, and a larger size of societal activity.

Reading the process of evolution in terms of complex systems theory, we observe that the drive toward instability is generated by the reciprocal influence between efficiency and adaptability operating on different time scales (a double asymmetry). The continuous transformation of efficiency into adaptability and of adaptability into efficiency pushes the system to the border of its basin of stability (to the edge of chaos). The steps of this circular cycle—with an arbitrary choice of the starting step—are: (1) an accumulation of experience in the system leads to more efficiency; (2) more efficiency increases the activity of self-organization by generating larger surpluses that can be invested in exploring new activities; (3) this increases the intensity and the scale of interaction between the socioeconomic system and environment, thus generating an increase in stress on the stability of boundary conditions and calling for an increase in adaptability; (4) in order to be able to invest more in adaptability, the system needs to be more efficient, that is, it has to gather experience on how to better deal with the present set of boundary conditions. At this point, the system returns to step (1).

This lends support to the idea that changes in boundary conditions are, sooner or later, inevitable for any evolving system. Indeed, these changes are implied by the very mechanism of evolution that, through Malthusian instability (Layzer 1988), follows the trend of maximum exergy degradation (Morowitz 1979; Jørgensen 1992; Schneider and Kay 1994) or Lotka's maximum power principle (H.T. Odum 1971, 1996).

The sustainability of complex systems can therefore only be imagined as a dynamic balance between efficiency (boosting the hypercycle by deleting obsolete activities) and adaptability (damping the hypercycle by introducing new activities). That is, the resonance between recipes and processes can be stabilized in time only by balancing short-term and long-term development. Survival can result only by a continuous change of structures to maintain functions, and a continuous change of functions to maintain structures. Put another way, neither a particular societal structure nor a particular societal function can be expected to be sustained indefinitely in the future. Practical solutions to this challenge depend on the nature of internal and external constraints faced by society. The definition and forecasting of these constraints is unavoidably affected by a large measure of uncertainty: Humans must continuously gamble trying to find a balance between efficiency and adaptability. In cultural terms, this means finding an equilibrium in focus between the past and the future of their civilization.

According to our model, a dissipative complex system evolves by increasingly integrating itself with its environment through a continuous enlargement of the scale and the intensity of its interaction. Therefore, we should describe this process as a process of the coevolution of society and its environment rather than a process of evolution of society within its environment (Salthe 1993; Norgaard 1994; Gowdy 1994). The parallel growth of the rate of energy dissipation (an intensive measure of the interaction between society and environment) and its area of interaction (an extensive measure of such interaction) means that in reality, what is growing along the third axis called SIZE is the product of the coevolution of the two interacting control systems, society and ecosystem, in their levels of compatibility and coordination.

Describing the Process of Evolution in Terms of Basins of Attraction

The model discussed in this paper describes the performance of human societies in their process of self-organization by using a set of biophysical indicators. Intensive variables, such as bioeconomic pressure (BEP), are used to assess the performance of society in terms of material standard of living, and environmental loading (EL) is used to assess the compatibility of the economic process with the biophysical processes responsible for the stability of the boundary conditions of society. An extensive variable, such as energy throughput (ET), can be added to the previous two to measure the size of society.

Clearly, we are not able to express in precise mathematical form how external and internal constraints affect the feasibility of the resonance between recipes and processes that stabilizes a social system at a certain dynamic equilibrium between BEP and SEH. On the other hand, we can express some of these parameters as functions of others and provide some indications of the nature of the mutual influence of intensive and extensive variables. We introduced the plane BEP-EL-SIZE (e.g., see Figures 4 and 19) to describe the evolution of society in terms of phase space. Such an analysis, according to insights provided by dynamic systems analysis and chaos theory, leads to the possible use of the concept of basins of attraction. Although we are presenting here only a working hypothesis, we feel that such a discussion has important theoretical implications for the description of society as an object moving within a quantifiable space of phases.

Three variables can be used, in a first approximation, to describe such a space of phases: (1) an intensive variable related to the level of dissipation, such as BEP, which is related to the exo/endo energy ratio. In this way, BEP is a function of the energy throughput (Watt) per unit of human mass (kg); (2) an intensive variable related to the coupling of the socioeconomic system with the environment, such as EL, which is related, at a particular level of technological development, to the energy consumed by society (the product of environmental impact of one unit of energy throughput and ET) and, thus, to the density of the human population in the ecosystem. In this way, EL is a function of human mass (kg) per area unit (m^2) of ecosystem; and (3) an extensive variable assessing the size of the self-organizing process.

The interaction of society/ecosystem can be estimated three ways: (1) in terms of total energy throughput (watts) of exosomatic energy controlled by society to stabilize a network of production and consumption activities; (2) total human population mass (kg) stabilized at a defined equilibrium between BEP and SEH with a certain set of technical coefficients; or (3) total area of ecosystem exploited by society (m^2) at a certain ET/m^2. The relationship among these three variables described in a plane $W/kg - kg/m^2$ (BEP-EL) makes the choice of any of these three parameters equivalent.

At this point, the process of evolution—or, better, coevolution, since the dynamic equilibrium between BEP and SEH (possible states accessible to the system) refers to characteristics of both society and ecosystem—can be described by the trajectory of society in the space BEP-EL-SIZE. In its trajectory of evolution, the system looks for combinations of these three variables that provide feasible dynamic equilibria, as illustrated in Figure 20. The three axes of this plane are:

y-axes = a parameter measuring the level of dissipation per unit of control (both values of W/kg and BEP are given);

x-axes = a parameter measuring environmental loading, W/m^2 (this is equivalent through W/kg to the use of the variable kg/m^2); and

z-axes = a parameter measuring the size of the system, total kg of population (this is equivalent through W/kg to the use of the variable total ET).

The use of the term basin of attractions in the space BEP-EL-SIZE is based on the idea that the probability of reaching a certain stable point of dissipation in the resonance "recipes ↔ processes" depends on the ability to lock the system of controls of the network of matter and energy flows in a solution BEP = SEH, at a certain size and environmental loading that falls

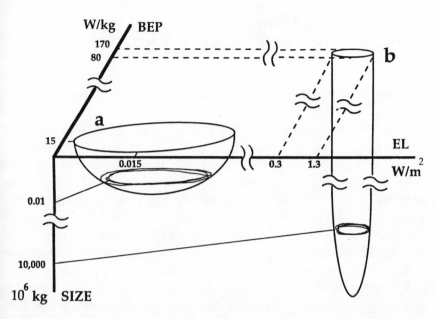

Figure 20.   Basins of attraction for socioeconomic systems in the space BEP-EL-SIZE.

within the range of values that is acceptable for the human compartment (BEP > critical BEP) and compatible with ecosystem characteristics (EL < Critical EL), and which the system of control is able to handle (compatible with available technology). The basin of attraction is generated by the ensemble of negative and positive feedbacks that keeps the system's characteristics within a range of values that enables the system of control to sustain in time the process of energy dissipation and vice versa. Within such a basin, the system of controls is able to adjust temporary imbalances between parameters generated by small perturbations, preventing their amplification to a scale that would induce the collapse of the dissipative system.

The surface of the basin is defined by values of critical environmental loading ratios and values of minimal acceptable BEP. A certain distance is kept by the societal system of controls between the set of states actually occupied by society and critical values on this surface. We saw earlier an example of safety buffers that limit the expansion of society generated by cultural control (e.g., pyramid building). Because of the need for this safety buffer, a higher BEP requires a higher FI/CI at the system level. The higher the BEP, the better the condition for lower-level holons (individual humans).

Mutual relations among parameters on different levels (on different time scales) generate the stability of the solutions BEP = SEH. These relations are not only determined by present conditions, but also by the history of the system. This is what Rosen (1991) calls "entailment." "The dictionary definition of entailment is a condition written in the original property title of rules of succession which is passed on from one generation to the next. Entailments in complex systems analysis are rules by which (future) events unfold from the system's structure and initial conditions" (Friend 1995, p. 12). What happened in the past through the building-up of experience (the accumulation of internal constraints) affects the present by reducing the options among the ensemble of paths accessible to the system (Brooks and Wiley 1988). Internal constraints (e.g., the experience stored in the system of control) reduce the choice among possible future paths of evolution and, therefore, project in front of the system a certain distribution of possible trajectories in the space of phases that is smaller than the distribution of trajectories possible according to biophysical constraints.[3] Anticipatory systems enable society to process information and run simulations about future scenarios. Therefore, human societies can affect the shape of their own basin of attraction in terms of the generation of distributions of conditional probabilities (about what is feasible) and by giving an arbitrary distribution of values to them (about what is auspicious) in terms of "passional entailment."

The dynamic behavior of socioeconomic systems is nonlinear due to the fact that all parameters determining BEP and all those determining SEH have to be changed together and in a coordinated way, because they affect each other (for a simple graphic representation of such a circular dependency among parameters, see Giampietro and Pimentel 1991). The need for a coordinated change of various parameters results in sets of forced relationships among these parameters, which in turn defines on the plane BEP-EL distinct areas of stability for the self-organization process of society. A socioeconomic system can either (1) remain within a particular basin of attraction, exploring the effects on lower-level holons induced by small adjustments in the value of parameters that stabilize the resonance between recipes and processes around the current equilibrium value of BEP = SEH; or (2) jump into a different basin of attraction, by trying a different combination of values able to match BEP and SEH, in a different space of the volume.

When society accumulates sufficient adaptability, some of the more active lower-level holons (e.g., social groups that are on the margin of society and are characterized by values of parameters distant from the societal averages) will break free from the attractor and abandon the old system of control (cultural identity, set of social rules, etc.). This process is the exact equivalent of speciation in biological evolution where marginal populations (groups that differ from the average characteristics of the species) break free from the genetic control of the species. The switch to another basin of attraction of a social system can be detected in the space BEP-EL-SIZE by a jump into a different basin of attraction (different set of values for the equilibrium BEP = SEH). This process of scanning for new possible solutions is enhanced when the situation of a socioeconomic system becomes unstable (when the whole social system has grown so big as to weaken the hypercycle ET/CI, or when some social groups become too smart or active to accept the burden of remaining under the old system of controls). In this case, the system breaks down into smaller pieces (lower-level holons) that start exploring the space around them in order to find new attractors. Tainter (1988) describes in this way the activities of local kingdoms after the collapse of the Roman Empire. Salthe (1993) describes this process as seedlings after the senescence of a complex system. When one of the "seedlings" manages to coordinate its activity with the activities of the environment and other societies into a new combination of variables that stabilize BEP = SEH into an encoded set of roles (rules and an effective system of controls), a new attractor is born. The new socioeconomic system can then grow (enlarge its scale to cut small perturbations) and develop.

Up to now, the trend in the evolution of human civilization has been that the more complex societies have had a basin that has one relatively small area on the BEP-EL plane, and another that is relatively deeper. The area of stability is collected at levels of higher dissipation (BEP) and larger interaction with the environment (EL), and the relative size is much higher (see numeric examples in Figure 20).

Different shapes of the basin of attraction observed for pre-industrial society (basin a) and industrial society (basin b), as shown in Figure 20, have several implications. Recalling the distinction proposed by Holling (1973) between resilience and stability, we find that basin a—the "small is beautiful" solution—is more resilient in the case of perturbation than basin b. Even if moved away from its present position on the plane BEP-EL, a pre-industrial society will be able to return to its previous trajectory due to the large and shallow form of the basin. The small size of the society also guarantees a lighter EL. On the other hand, the large size of a more complex system represented by basin b—the "big is better" solution—implies a stronger stability of the system, since it is more difficult to bring the system outside its basin. This stability is paid for by a greater fragility of the system in the case where perturbations are able to move it away from its present position. Greater fragility is explained by the smaller area of stability on the plane BEP-EL. For example, in the case of a cut in the oil supply, Manhattan would be more fragile than a small village of farmers in Bolivia. On the other hand, cutting the oil supply of Manhattan is much more difficult than cutting the oil supply of a small village in rural Bolivia. (Figure 20 was originally prepared by the first author to describe on the same set of variables different types of ecosystems, with a tundra being type a basin and tropical forests being type b basin.)

The two basins illustrated in Figure 20 represent possible combinations of BEP, EL, and Size that guarantee the stability of the attractor. Large resilience—that is, the "small is beautiful" solution (a)—is paid for by the system in two ways: (1) high risk at the level of the individual, because the system absorbs perturbations at the lower level. Such a system adjusts its internal parameters (the ones determining BEP) when SEH is decreased; and (2) low specialization, for the small size of the system does not allow a large diversity of behavior. Large stability, the "big is better" solution (b), is paid for two ways: (1) high risk at the level of society, because the structure of society is extremely fragile if and when the system is unable to prevent changes in boundary conditions. In fact, the system absorbs perturbations at the focus level using its large scale to overcome local perturbations, rather than at the individual level. For example, almost nobody in Manhattan would be able to start a fire without matches; and (2) high environmental loading ratio, which translates into the need to undergo continuous processes of change at increasing speed. This forces large systems to invest heavily in adaptability due to the large probability of changes in boundary conditions.

Finally, the two different forms of the basin of attraction imply larger or smaller fluctuations in the values of parameters and longer or shorter lag times in regulation.[4] According to our model, this implies that in preindustrial societies the control of the stability of the attractor is obtained through wide fluctuations in the parameters related to the human compartment (population density and physiological variables), whereas in industrial societies fluctuations are observed in the exosomatic compartment (economic cycles of investment and disinvestment and fluctuations in debt and national deficit), which is much better for the lower-level holons—the individual humans—who live in it.

This is one of the main reasons that explains the strong drive toward unsustainability experienced today by humankind. The "big is better" solution has found several ways to escape (depleting stocks, filling sinks, trading, incurring debt), at least temporarily, the constraints that would result from a deeper respect for ecological equilibria. In doing so, developed societies have managed to provide to their inhabitants a material standard of living much higher than that typical of pre-industrial societies.

Even though humans have begun to understand that the current matching of BEP and SEH achieved by Western civilization is not sustainable in the long term, very few are willing to settle for a lower level of BEP just for the sake of increasing the "life span" of present boundary conditions. After all, present boundary conditions will sooner or later change anyway. This leaves us with a huge set of ethical questions related to the choice of a path for the future development of humankind. The terms of the dilemma are: (1) wide differences in levels of material standard of living among and within countries, (2) uncertainty regarding the resilience of our biosphere to human-generated stress and the ability of human ingenuity to cope with shortages of resources, and (3) differences in cultural definitions of what is an auspicious path of development for socioeconomic systems.

The positive contributions that the model presented in this paper can provide to the debate are: a general clarification of the terms of the problem by enabling a bridge among different perspectives, and an explicit acknowledgment of the irreducible nature of conflicts in complex systems and the innate unpredictability of their behavior. Because of the complexity of the process of interaction between socioeconomic systems and environments, the problem of sustainability does not have one unique and "best" solution. It simply cannot have just one. There is always a legitimate plurality of perspectives that cannot be merged into a unique cost/benefit analysis, no matter how smart the researcher performing the analysis or how big the computer running the model. The "optimal path for sustainable development" is a myth. Certainly, scientists can provide a contribution within these limitations, but only if we stop pretending that these limitations do not exist.

ACKNOWLEDGMENTS

The authors wish to acknowledge the helpful advice provided by Eugene A. Rosa, Lee Freese, and C. Dyke for both parts of this work.

NOTES

1.   When a hierarchical frame of analysis is adopted, we find that obligations toward future human generations are the same as the obligations we have toward other species and

biota living together with the present human generation. When enlarging the spatiotemporal scale of assessment, it is impossible to define different obligations to and rights of humans and other species, since they are all part of the same hierarchy. Preserving the stability of the biosphere today means preserving the right of human generations in the future (Giampietro 1994a).

2. Clearly, inflation played a role in the dramatic increase in per capita GNP. However, the phenomenon of inflation can be seen as an increase in the resolution at which economic systems, after a perturbation, search for comparative advantages among different activities. During a period of hyperinflation, obsolete activities and roles are abandoned more rapidly. In this way, the system quickly creates a picture of the "new" boundary conditions by relocating power in its system of controls.

3. The definition of information given by Shannon: "*information is anything that causes an adjustement in a set of probabilities assigned to a set of possibilities*" (cited in Tribus and McIrvine 1971, p. 179) should be recalled here.

4. This issue is as fascinating as it is complicated. Watt (1989, 1992) proposes that different lag times in the adjustment of parameters determining the stability of these attractors can be used to explain the existence of wave behavior in socioeconomic systems, such as Kondratieff's economic cycles. Watt's analysis indicates several lag times that induce a systemic behavioral response in the evolution of societies: (1) short-term business cycles: change in dimension of stocks of goods; (2) short/medium term cycles: investment cycle changes in the B&M loop; (3) medium-term cycles: demographic changes reallocating population mass among age classes; and (4) long-term cycles related to fundamental shifts in the choice of energy sources, that is, changes in the nature of the process of exergy dissipation fueling the hypercycle. Watt's findings are perfectly consistent with the schematization provided by our model.

REFERENCES

Abernethy, V. 1991. "Editorial: How Julian Simon Could Win the Bet and Still be Wrong." *Population and Environment* 13(1): 3-7.

Brooks, D.R., and E.O. Wiley. 1988. *Evolution as Entropy.* Chicago: The University of Chicago Press.

Brown, M.T., P. Green, A. Gonzales, and J. Venegas. 1992. *Emergy Analysis Perspectives, Public Policy Options and Development Guidelines for the Coastal Zone of Nayarit Mexico.* Report to the Custeau Society and the Government of Nayarit Mexico. Gainesville: Center for Wetlands and Water Resources, University of Florida.

Cherfas, J. 1991. "Skeptics and Visionaries Examine Energy Savings." *Science* 251: 154-156.

Cottrell, W.F. 1955. *Energy and Society: The Relation Between Energy, Social Change and Economic Development.* New York: McGraw-Hill.

Debeir, J-C., J-P. Deléage, and D. Hémery. 1991. *In the Servitude of Power: Energy and Civilization through the Ages.* Atlantic Highlands, NJ: Zed Books Ltd.

Ehrlich, P.R., A.H. Ehrlich, and G.C. Daily. 1993. "Food Security, Population, and Environment." *Population and Development Review* 19(1): 1-32.

Friend, A.M. 1995. "Sustainable Development Indicators: A Conservation Accounting Approach." Background Paper presented at the ISEE/RC Workshop on "Socio-Ecological Economic Systems: From Information and Simulation to Practical Solutions." Perislav-Zalesski, Russian Federation, July 16-20.

Funtowicz, S.O., and J.R. Ravetz. 1994. "The Worth of a Songbird: Ecological Economics as a Post-normal Science." *Ecological Economics* 10(3): 197-207.

Gever, J., R. Kaufmann, D. Skole, and C. Vörösmarty. 1991. *Beyond Oil: The Threat to Food and Fuel in the Coming Decades*. Niwot: University of Colorado Press.

Giampietro, M. 1994a. "Sustainability and Technological Development in Agriculture: A Critical Appraisal of Genetic Engineering." *BioScience* 44(10): 677-689.

_____. 1994b. "Using Hierarchy Theory to Explore the Concept of Sustainable Development." *Futures* 26(6): 616-625.

_____. 1997. "Linking Technology, Natural Resources, and the Socioeconomic Structure of Human Society: A Theoretical Model." Pp. 75-130 in *Advances in Human Ecology*, Vol. 6, edited by L. Freese. Greenwich, CT: JAI Press.

_____. In Press. "Demographic Pressure, Socio-Economic Pressure and Technical Development of Agriculture." *Agriculture, Ecosystems and Environment*.

Giampietro, M., S.G.F. Bukkens, and D. Pimentel. 1993. "Labor Productivity: A Biophysical Definition and Assessment." *Human Ecology* 21(3): 229-260.

Giampietro, M., and S.G.F. Bukkens. 1996. "Energy Budget and Demographic Changes in Socioeconmic Systems." Paper presented at the Inaugural Conference of the European Branch of the International Society for Ecological Economics: "Ecology, Society, Economy: In Pursuit of Sustainable Development." Université de Versailles, Saint Quentin en Yvelines (UVSQ), France, May 23-25.

Giampietro, M., and K. Mayumi. 1996. "Carrying Capacity and Optimum Population: Theoretical definitions and Simulations." Paper presented at the Fourth Biennial Meeting of the International Society for Ecological Economics: "Designing Sustainability," Boston University, August 4-7.

Giampietro, M., and D. Pimentel. 1991. "Energy Efficiency: Assessing the Interaction Between Humans and Their Environment." *Ecological Economics* 4: 117-144.

Giampietro, M., S. Ulgiati, and D. Pimentel. In Press. "Biofuel Production as an Alternative to Oil: A Reality Check." *BioScience*.

Gowdy, J.M. 1994. *Coevolutionary Economics: The Economy, Society and the Environment*. Boston: Kluwer Academic Publishers.

Hall, C.A.S., C.J. Cleveland, and R. Kaufmann. 1986. *Energy and Resource Quality*. New York: John Wiley & Sons.

Holling, C.S. 1973. "Resilience and Stability of Ecological Eystems." *Annual Review of Ecological Systems* 4: 1-23.

ISTAT. 1992. *Annuario Statistico Italiano*. Rome: Istituto Centrale di Statistica.

James, W.P.T., and E.C. Schofield. 1990. *Human Energy Requirements*. Oxford: Oxford University Press.

Jevons, F. 1990. "Greenhouse—A Paradox." *Search* 21 (5): 171-172.

Jevons, W.S. 1865. *The Coal Question*. London: Macmillan.

Jørgensen, S.E. 1992. *Integration of Ecosystem Theories: A Pattern*. Dordrecht: Kluwer Academic Publishers.

Kaufmann, R.K. 1992. "A Biophysical Analysis of the Energy/Real GDP ratio: Implications for Substitution and Technical Change." *Ecological Economics* 6 (1): 35-56.

Kendall, H., and D. Pimentel. 1994. "Constraints to the World Food Supply." *AMBIO* 23: 198-205.

Khazzoom, J.D. 1987. "Energy Saving Resulting From the Adoption of More Efficient Appliances." *Energy Journal* 8 (4): 85-89.

Layzer, D. 1988. "Growth of Order in the Universe." Pp. 23-40 in B.H. Weber, D.J. Depew and J.D. Smith, eds., *Entropy, Information, and Evolution.* Cambridge, MA: MIT Press.

Meadows, D. 1994. "Envisioning Sustainable Alternatives." Plenary lecture given at the Third Biennial Meeting of the International Society for Ecological Economics *"Down to Earth,"* October 24-28, San José, Costa Rica.

Mendelsshon, K. 1974. *The Riddle of the Pyramids.* London: Thames and Hudson.

Morowitz, H.J. 1979. *Energy Flow in Biology.* Woodbridge, CT: Ox Bow Press.

Newman, P. 1991. "Greenhouse, Oil and Cities." *Futures* (May): 335-348.

Norgaard, R.B. 1994. *Development Betrayed: The End of Progress and a Coevolutionary Revisioning of the Future.* New York: Routledge, London.

Odum, E.P. 1971. *Fundamentals of Ecology.* Philadelphia: Saunders.

Odum, H.T. 1971. *Environment, Power, and Society.* New York: Wiley-Interscience.

_____. 1996. *Environmental Accounting, Emergy and Decision Making.* Gainesville: University of Florida Press.

Pastore, G., M. Giampietro, and K. Mayumi. 1996. "Bio-Economic Pressure as Indicator of Material Standard of Living." Paper presented at the Fourth Biennial Meeting of the International Society for Ecological Economics: "Designing Sustainability," Boston University, August 4-7.

Pimentel, D., and M. Pimentel. 1979. *Food, Energy, and Society.* London: Edward Arnold.

Rosen, R. 1991. *Life Itself: A Comprehensive Inquiry into the Nature, Origin and Fabrication of Life.* New York: Columbia University Press

Salthe, S.N. 1993. *Development and Evolution: Complexity and Change in Biology.* Cambridge, MA: The MIT Press.

Schneider, E.D., and J.J. Kay. 1994. "Life as a Manifestation of the Second Law of Thermodynamics." *Mathematical Computer Modelling* 19: 25-48.

Simon, J.L. 1981. *The Ultimate Resource.* Princeton, NJ: Princeton University Press.

Tainter, J.A. 1988. *The Collapse of Complex Societies.* Cambridge: Cambridge University Press.

Tribus, M., and E.C. McIrvine. 1971. "Energy and Information." *Scientific American* 224: 179-189.

United States, Bureau of the Census. 1990. *Statistical Abstract of the United States, 1990.* Washington, DC: U.S. Department of Commerce.

_____. 1991. *Statistical Abstract of the United States, 1991.* Washington, DC: U.S. Department of Commerce.

Watt, K. 1989. "Evidence for the Role of Energy Resources in Producing Long Waves in the United States Economy." *Ecological Economics* 1: 181-195,

_____. 1992. *Taming the Future.* Davis, CA: The Contextured Web Press.

Weber, J. 1996. "Evaluation of Nature: Nature of Values." Paper presented at the Inaugural Conference of the European Branch of the International Society for Ecological Economics, Université de Versailles Saint Quentin en Yveline, Paris, France, May 23-25.

World Resources Institute. 1994. *World Resources 1994-95.* New York: Oxford University Press.

SURPLUS AND SURVIVAL:
RISK, RUIN, AND LUXURY IN THE EVOLUTION
OF EARLY FORMS OF SUBSISTENCE

Ulrich Müller-Herold and Rolf P. Sieferle

ABSTRACT

In their search for "primitive man," anthropologists of the 1970s constructed an "original affluent society" in the Stone Age, when people led a
life of leisure and relative superabundance embedded in the harmonies of
nature. These early subsistence economies have been reconstructed as based
on principles of risk minimization and leisure preference. Here, we show
that risk minimization is just one element in a wider spectrum of coping
with risks and the danger of ruin. In subsistence economies, long-term survival requires a broad spectrum of different behavioral patterns. In situations of scarcity or danger, the acceptance of calculated risks can increase
the chances of survival per saltum. Under less extreme conditions, risk prevention by portfolio formation is generally more effective as a strategy to
reduce fluctuations. Risk prevention can finally be made more profitable by
suitable technological management of fluctuations. Thus, a risk spiral as a

Advances in Human Ecology, Volume 6, pages 201-220.
Copyright © 1997 by JAI Press Inc.
ISBN: 0-7623-0257-7

dynamizing principle in the development of complex societies is generated: The reduction of a particular risk leads to new types of uncertainty, which in turn require further (risky) innovations. This mechanism creates a permanent innovation pressure responsible for the restless transformations in complex societies.

ACCOUNTS OF THE NATURAL STATE

Ever since its earliest days, European thinking has posited two distinct but nevertheless complementary accounts of the origins and the course of history (Lovejoy and Boas 1935). The one tells of an original, paradisiac state, of a Golden Age in which human beings free of sin lived in superabundance, peace, and harmony with both nature and their own kind. History is thus understood as a process of estrangement from this origin, either by virtue of a disastrous breach, an expulsion from paradise, or else through humanity's experience of irreversibly deteriorating conditions. By contrast, the complementary basic narrative tells of a natural state in which, according to Thomas Hobbes' classic formulation, the life of man was solitary, poor, nasty, brutish, and short (Hobbes 1651). Following this premise, things can only get better in the course of history, with the result that the task of the future is to provide for institutionalized security in a civil society as well as to increase production and control the natural world.

Each of these notions of history is, of course, a product of the particular present which gave it birth. Viewed as the result of an historical process, however, this "present" in a certain sense is the culminating point of a movement that has strayed away from its origins. The characteristics of the present are, therefore, at odds with those of its origins, which is to say that the respective present constitutes the opposite of the natural state, with the result that it displays in reversed form all those characteristics which its constructor misses in the present. The original state is, as it were, the end turned on its head.

This structure is most clearly apparent in the figure of "primitive man." Even in the ancient world, the barbarian and the noble savage were the basic forms in which the difference between a foreign, retarded culture and the observer's own culture could be expressed. To anyone who ascribes a high standing to his or her own culture, the alien appears to be a barbarian, an ill-mannered brute incapable even of proper speech. Looked at the other way round, however, the noble savage can also constitute an idealized negative image that permits criticism of the existing state of affairs. Tacitus, for instance, described the Germani as bold and highly moral warriors—a depiction intended to hold up a mirror to his Roman contemporaries, who lacked these qualities. The savage is poor or modest, threatened by death or courageous, boorish or graceful—all according to the standards that have been set by the perceiver's own culture.

The question as to which individual characteristics of the noble savage are regarded as enviable is thus something to be decided by the observer. Tacitus, for example, who reprimanded the decline of morals in Rome, praised the chastity of the Germani. In the 19th century, by contrast, the evolutionist Lewis Henry Morgan, to cite one instance, regarded savages as promiscuous, living in a "wenching society" that excited the horror, but also the secret fantasies, of the Victorians (Morgan 1877). In the early 20th century, Margaret Mead, for instance, was then able to posit the free and natural sexuality of the Polynesians as exemplary by comparison with the inhibitedness of civilization (Mead 1928). In the closing years of the 20th century, an attitude of scepticism has developed vis-à-vis such myths (cf. especially Duerr 1989). Only for the feminist suffering under the "patriarchy" of the present does the noble savage exist, in the form of an original "matriarchy" from which people have become fatally estranged and to which they should return. The fact that such a matriarchy is at all realizable, however, can be proven by its existence in a distant past.

IN THE BEGINNING WAS SUPERABUNDANCE

For a long time, agreement existed that the process of "progress," of "civilization," of "modernization," or of "economic growth" had produced at least one unequivocal result, namely, that of increasing material affluence. It was, indeed, still possible to connect this increased affluence with a soulless way of living, an existence gone wrong, but there could be no doubting the increase in affluence as such. In this view of things, technical progress and economic growth constituted in a certain sense the hard core of the historical process, one which—whatever value might be placed upon it—was unquestionably real. Chaste or sensual, gentle or cruel, bellicose or pacific as the savage may have been, he certainly did not live in luxury; rather, his existence most certainly bore the stamp of "the simple life."

This view was seriously undermined when in 1972, the American anthropologist Marshall Sahlins (1974) characterized Stone Age society as the "original affluent society." Drawing on field studies, above all on Richard Lee's studies of the bushmen of the Kalahari (Lee 1968, 1979), Sahlins drew a picture of primitive societies which, for the modern observer, bears striking similarities to the layabout utopia of the hippie movement: work and discipline are unknown, but there is still no lack of anything, as people's requirements are well below the material potentialities, and they possess above all a superabundance of that which is prized most highly: free time for playing and communicating with each other.

Anthropologists of earlier generations had always mistakenly pictured Stone Age people as living on the brink of starvation. According to this view, their lives consisted—as Marvin Harris ironically put it in retrospect—of a continuous battle for survival, "a time of great fear and insecurity, when people spent their days ceaselessly searching for food and their nights huddled about fires in comfortless caves besieged by saber-toothed tigers" (Harris 1978). Sahlins now demonstrated that this picture by no means corresponded to reality. The research done by Lee, in particular, showed that even in such a hostile environment as that of the Kalahari desert, a life of relative material affluence was possible. People had an adequate, balanced diet, consisting of one-third meat and two-thirds plants, with a high proportion of protein. The women gathered over a hundred different species of roots, nuts, and tubers, spending not more than two or three days on the process. They did not build up supplies of food, as sufficient quantities were always available for gathering. The men likewise dedicated two to three days a week to the hunt, pursuing not less than 54 edible species of animal. The results here were less predictable than in the case of gathering but, by way of compensation, success in the hunt enjoyed higher social prestige, especially when a rare and tasty quarry was caught.

The general state of health in this hunter-gatherer society was usually described as good (a rather more differentiated conclusion is reached by the comprehensive literature study of Cohen 1989). Malnutrition and deficiency diseases were as rare as obesity or diabetes, nor were there any infectious diseases such as cholera, typhus, measles, smallpox, or influenza, nor caries, appendicitis, or cases of cancer. It must be said, however, that life expectancy was considerably lower than that in modern industrial societies, even though it was higher than in traditional agrarian societies, in which the high density of population facilitated the spread of infectious diseases. People were helpless in the face of accidents and injuries, septic wounds, and parasites, and infant mortality was also high. What especially attracted the attention of the researchers, however, was the fact that people spent an average of no more than four hours a day on "work," that is to say, on hunting, gathering, and preparing food. Truly the exact opposite of late capitalist industrial society, marked as it is by consumerism and pressure to perform!

SUBSISTENCE ECONOMIES AND THE MINIMIZING OF RISK

One would be doing Sahlins an injustice, however, if one were to reduce his reflections simply to the construction of a utopia of drop-outs, even if these connotations resonate in his works and make plausible the broad reception of this view of history during the seventies. Sahlin's work and the literature that followed in the wake of his study did not merely seek the contemplative "noble savage" marked by postmaterialistic values; rather, it genuinely aimed at the reconstruction of the functional significance of this behavioral pattern, as finally presented in a coherent manner by Dieter Groh (1992). Groh is concerned with the general reconstruction of the social logic of "subsistence economies," which is to be categorically distinguished from that of other economies—above all, market economies. Subsistence economies existed among Palaeolithic hunter-gatherer societies as well as among early peasant societies. It is only the agrarian civilizations that developed about some 5,000 years ago that differ from this pattern, although in these, too, there are subsistence economy elements which extend even into the European societies of the modern age.

The fundamental distinction between a subsistence economy and its opposite and historical successor, the market economy, resides in the fact that the latter is orientated around the production of surplus, or on the principle of the maximization of yield. The basic strategy attributed to subsistence economies, by contrast, is that of "minimization of risk," by which Groh means that it aims at the stationary occupation of a niche in a particular habitat and, thus, eschews the taking of certain opportunities which, perhaps, are not rewarded in the longer term.

Now, Groh assumes that the identification of such a "strategy" involves an outside observer's construct, which is usually not identical with the inside view of the society in question. If one asks a member of a subsistence economy why he does not, despite the usual availability of resources, increase his "production," the member will surely not answer by referring to a strategy of risk minimization. For this reason, the representatives of such a society do not necessarily "know" what is going on when they practice "risk minimization." Their "strategy," rather, internally assumes forms which relate to their purpose in a purely functional manner but which are not accessible to conscious representation.

Within its social logic, the basic model of a subsistence economy thus distinguishes between two elements that on the one hand represent the functional exterior view and on the other the symbolized interior view of the respective society. The strategy of "risk minimization" thus expresses itself in a unity of "underproductivity"—by which is meant the systematic under-use of resources—and a behavioral orientation in the sense of "leisure preference."

Leisure preference—or laziness, in common parlance—is a form of behavior in primitive societies that at a very early stage attracted the (disparaging) notice of observers coming from a commercial or industrial society. Colonizers repeatedly complained about the behavior of the indigenous or local lower classes of society, which consisted of ceasing to work as soon as they thought they had earned enough. This was a mode of behavior that at first sight appeared to run entirely counter to the logic of the modern market economy. In a market economy, one would expect the supply of a commodity to rise with the demand for it, that is, with the price to be attained. If wages are high, then the supply of work must increase, which means that workers must be willing to put in well-paid overtime. In the context of subsistence economies, an entirely contrary way of acting was observed. When wages are high, less work is done, with people working only as many days as necessary to earn a wage regarded as appropriate. A further increase of wages intended as an incentive only leads to a further reduction in hours worked. It is this mode of behavior that is termed the leisure preference: The freedom to dispose of one's time as one wishes is valued more highly than are the goods which could be purchased with increased wages. Immaterial leisure is prized more highly than material income.

In the context of a market economy, such behavior need not count as imprudent if it can be interpreted to the effect that the individual worker perceives a greater benefit in the avoidance of "the misery of work" than in the increase of the consumption of material goods. The logic of a *Homo oeconomicus* maximizing benefits would thus not be shattered by the observation of leisure preference. Such a pattern of preference, however, is regarded as pretty unusual in modern society, and before talk of a "post-materialist transformation of values," it carried a whiff of laziness and antisocial tendencies. This may be due to the fact that within a social system orientated toward dynamism and economic growth, the expansion of such a pattern of preference would not be of much help and would tend to undermine the system.

Within the basic strategy of a subsistence economy, by contrast, leisure preference expresses no imprudence, no lack of discipline or virtue, but also no noble rejection of base materialism; rather, a functional orientation in the sense of a specific adaptation to certain environmental conditions can be accorded to it. For "underproductivity," or under-use of resources associated with leisure preference, constitutes an evolutionarily successful strategy if the adaptation resides less in the attempt to actively control access to resources than in the process of partaking of an already existing flow of resources. Such a strategy is not brimming with progressive energy, but it has shown itself to be successful insofar as it has allowed the survival of humanity over extremely long periods of time.

BEYOND RISK MINIMIZATION

Closer inspection, however, reveals the picture of subsistence economies sketched here as requiring additions and corrections. The indiscriminate use of the term risk minimization for subsistence economies is in itself problematic in that it implies that risks are avoided at all costs. It can be shown, however, that in particular situations of scarcity or danger, it is eminently sensible—and also an observable fact—to accept calculated risks if, by this means, the chances of survival are increased per saltum. A case of this nature has been presented by Ruth Mace (1993), using the Gabbra, a pastoral people of East Africa.

The Gabbra inhabit an area of low and unpredictable rainfall, where they subsist from the husbandry of sheep, goats, and camels. Rainfall is too low to support agriculture. Studies of nearby Turkana pastoralists suggest that the herding system is ecologically sustainable. The major threat to the herds is drought. The herders are well aware that breeding females, in particular, are highly endangered, lactation being the costliest part of reproduction for mothers. This leads to the herders' basic dilemma: On the one side, a herder's wealth relates to the number of offspring; breeding females, on the other, constitute the dominant risk for the herd's survival.

Mace demonstrates that herders forego short-term gain in favor of long-term household survival. This is done by the herders' sometimes reducing the breeding rate of herds by controlling the access of males to females. The practice is associated with wealth and is thus prestigious. Herds where the breeding rate was uncontrolled are described as "getting finished very quickly." Despite appreciating these facts, herders with small numbers of sheep choose not to control the breeding rate of their sheep, claiming they "need more animals." The owners of larger herds reduce the risk by reducing the number of breeding females, reducing short-term gain at the same time. In contrast, herders with small numbers of sheep, facing the danger of immediate loss of the last animals, choose the riskier option of unrestricted reproduction. (In line with risk research, a type of behavior is defined as risk-prone when, given the choice between two alternatives, one alternative with greater insecurity in its consequences is consistently selected.)

Mace's study revealed that, in view of a greatly fluctuating environmental situation, two complementary modes of behavior were to be observed within the same population. Anyone in the relatively safe position of having a large herd does, indeed, attempt to minimize risks. Anyone whose herd is too small to cope with fluctuations, however, puts everything on a single card, so to speak, as in this way there is at least a chance that the herd might manage to survive. Poor and wealthy households, accordingly, adopt different behaviors which can be seen as elements of a superior strategy whose objective is to guarantee the long-term existence of a group of families in such a manner that the shortage of a limiting resource does not lead to destruction. It turned out, in addition, that individual households adapt to changes in size: At a well-defined turning point, they switch from risk-avoiding to risk-prone behavior if the herd declines, and vice versa. Corresponding strategies can also be demonstrated in the animal world (Caraco et al. 1980).

This superior overall strategy of maximizing long-term household survival should be termed a "strategy of maximum ruin avoidance:" It can be regarded as a kind of Gambler's Ruin game, in which players have an initial fortune (Slobodkin and Rapoport 1974). If the players agree in advance of starting the game that a player will die if that player looses his fortune completely, then there is no question of the player cashing in his chips, that is, there is no external payoff. A player in this situation will play only to maximize the probability of persisting in the game or, stated in complementary terms, to minimize his probability of ruin. This goal of ruin avoidance is highly similar to what the Gabbra are trying to achieve with their herds. Decisive for the evaluation of a strategy of ruin avoidance as one of "maximum ruin avoidance" is the switch from risk-prone to risk-avoiding behavior at a theoretically well-defined turning point. Mace was able to demonstrate that the turning point observed in the Gabbras was about the same as the theoretical turning point calculated from a suitable dynamic programming model.

In subsistence economies, "maximum ruin avoidance" describes a general, comprehensive overall strategic frame within which risk minimization is only one—albeit frequently occurring—element among others. Closer examination of risk minimization can ultimately demonstrate that the combination of underproductivity or under-use of resources with leisure pref-

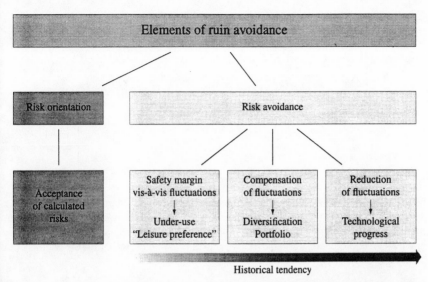

Figure 1. Elements of ruin avoidance. In subsistence economies, long-term survival requires a broad spectrum of different behavioral patterns. In situations of scarcity or danger, the acceptance of calculated risks can increase the chances of survival per saltum. Under less extreme conditions, risk avoidance is generically more effective as a strategy. Risk-avoiding behavior can be made more profitable economically if it is complemented by suitable mangement of fluctuations.

erence once again represents merely a certain band of a larger spectrum of behavior. Within the spectrum, historically, solutions in the sense of a port-folio formation have developed, which have ultimately led to the creation of a dynamic process in the sense of a "risk spiral" (see Figure 1).

STRATEGIES OF RUIN AVOIDANCE

In the following, an attempt will be made to examine various strategies of subsistence economies in the historical process. This is a reconstruction in the sense of conjectural history, which is to say that the individual stages differentiated here are primarily logical rather than chronological in nature.

When the availability of resources in a particular habitat fluctuates and a group wishes passively to adapt to these fluctuations, a prudent strategy is to stabilize consumption below the effective lower margin of this fluctua-tion. As shown in Figure 2, we have a certain human habitat within which a limiting factor, r—for example, water, game, or certain edible plants—fluctuates at irregular intervals. In this first and most simple instance, which probably corresponds to the mode of existence of Palaeolithic

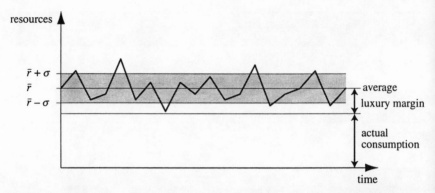

Figure 2. Under-use. When the population size is limited, demand remains so low that in the case of major fluctuations in a downward direction—even greater than the standard deviation σ—the minimum subsistence level is guaranteed. On average, however, resources are available on a scale that exceeds actual consumption by a luxury margin.

hunter-gatherer societies, the group orientates itself around the lower fluctuation margin of the limiting factor that is to be anticipated in the long term. By this means, it maintains a safety margin to the average value. It follows that this society almost always lives in circumstances that can be described as superabundance. This luxury margin can be very far removed from the lower fluctuation limit, but the latter constitutes precisely the bottleneck through which, in some extreme situation, society may have to pass. A growth in population or of the material throughput into a superior margin would in the longer term be (ceteris paribus) fatal.

The greater the fluctuation (standard deviation), the greater is the expected superabundance under normal conditions. Richard Lee, for instance, estimates that the !Kung San in the Kalahari generally use only half the average available food resources—and these are normally those that can be procured with the least effort. In cultural terms, this behavioral pattern finds expression in the observed leisure preference, which functionally orients the individuals toward the under-use of resources. The ecological effect of this strategy is the sustained occupation of a niche that places little strain on the environment. This strategy has its price, however: society has actively to control its fertility, as the population size is not controlled in Malthusian terms by the scarcity of environmental factors. For this reason, primitive societies have developed diverse methods of cultural procreation control, with infanticide as the ultima ratio (Harris and Ross 1987).

It can be shown, however, that not only the orientation at the lower edge of the fluctuation spectrum but also other behavioral patterns can be under-

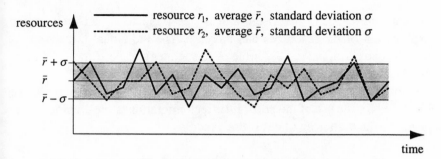

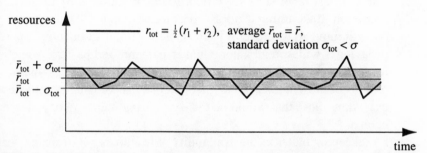

Figure 3. Compensation of fluctuations by means of portfolio formation. A limiting factor is assumed to fluctuate in two different habitats. In both habitats, the average yield of the respective resource is r and the fluctuation (standard deviation) σ. A reduction of the overall fluctuation σ_{tot} comes about thus: The variance of the portfolio is

$$4\sigma_{tot}^2 = (\sigma_1^2 + \sigma_2^2 + \sigma_1\sigma_2\rho) = 2\sigma^2(1+\rho)$$

ρ here characterizes the correlation between the yield of both resources. ρ is bigger than -1 and smaller than 1. (The second equals sign comes from the fact that the variances in both segments are assumed to be of equal size: $\sigma_1 = \sigma_2 = \sigma$.) For the fluctuation of the portfolio as a whole, the following thus applies:

$$\sigma_{tot} = \sigma\sqrt{(1+\rho)/2} < \sigma \text{ if } \sigma < 1$$

The fluctuation of the portfolio is thus smaller than the fluctuation in the segments if and only if the correlation is smaller than 1. (This excludes portfolio effects by means of trivial apportionment.)

stood as successful adaptations in the sense of ruin avoidance—at least for short and middling periods of several thousand years. For this purpose, we take the following basic model as our starting point. In a particular habitat, a limiting factor, r, fluctuates in time. We assume that two different habitats, 1 and 2, in which the limiting factors r_1 and r_2 fluctuate (more or less) independently of one another, are used with half the intensity. Their combined use is then subject to a lesser fluctuation, and the lower edge of the fluctuation band comes close to the mean value (see Figure 3). In this case, the carrying capacity has risen, that is, the population can grow or the exploitation of the resources can increase. The individual fluctuation as such has not disappeared by virtue of this strategy of portfolio formation, but its negative effect is partially or completely compensated.

In historical terms, four different strategies of diversification can be identified for the use of the portfolio effect: migration, storekeeping, transport, and the scattering of cultivable land:

1. The simplest method of spatial portfolio formation is by means of migration: If a limiting factor has become scarce in one habitat, people will move to wherever it is available in sufficient quantity. Such behavior is widespread among hunter-gatherer and pastoral societies, and it contains no incentive for the development of a dynamic process because it leads to no interventions in ecological complexes exceeding those that would be the case during mere use of a single habitat.

2. Storekeeping balances the availability of resources not in spatial but in temporal terms. The actual availability of a limiting factor loses importance when this factor is still available from a previous surplus situation. The storing of supplies is thus to be seen as a strategy of portfolio formation which brings the actual availability of resources nearer to the long-term average. It must be said, however, that storekeeping does not merely require the existence of foodstuffs which (like grain) can be stored but, more than this, it demands continuous control of the stored supplies, which is only possible for people leading a sedentary form of existence. For this reason, efficient and regular storage of supplies does not occur until the advent of the agrarian mode of production.

3. If migration means that people move to resources, then of course resources can also be brought to the people. This is transport by (long-distance) trade. To be efficient, however, it requires a host of technological methods, such as the availability of suitable means and routes of transport, motive power, and the like. Under primitive conditions, it is limited to a small number of luxury items such as amber or seashells. Trade of foodstuffs on a larger scale can only be found under the conditions of agrarian civilizations. Here, however, trade gains the function of creating an average of resource availability. The greater the distances involved and the more varied the areas in ecological and climatic terms, the greater is the portfolio effect.

4. Finally, a fourth kind of portfolio formation is the diversification of cultivable areas, which is mainly to be found in agrarian civilizations based on seed crops. A well-researched example of this is the traditional open-field system of English agriculture in the medieval and early modern ages, which can be found in similar form in other agrarian societies as well. In this model, a single peasant household owns a variety of small parcels of land which are distributed across the entire area of the village land. The effects of damage to local fields by hail, fire, drought, wind, flooding, and hunting can be kept down by this means, but these advantages have to be paid for with specific disadvantages (increased distances between the parcels of land, difficulty of improving land, enforced consensus between villagers obliged to work the same land).

A good example of this strategy is presented in McCloskey's attempt to reconstruct the English open-field system as a strategy of risk minimization (McCloskey 1976; Cashdan 1990): In the Middle Ages, yields from agriculture fluctuated considerably. In order to make up for them as well as possible, the portfolio effect was used in two ways, on the one hand, by the combined cultivation of various types of grain (wheat, oats, and barley),[1] and on the other by the division of the growing areas for one and the same variety of grain into smaller-sized fields at places separated from one another on the common land.[2] The optimum division on the part of each peasant of the cultivable land into 6-10 individual fields led to a total yield that was some 10 percent lower than in the case of fields that were not divided up. The frequency of famine, however, was reduced by this process from an average of once every nine years (in the case of fields that were not divided up) to an average of once every 14 years. A slight reduction in the harvest was thus set against a considerable increase in the security of the food supply. The success of the portfolio strategies was so great that the process of putting the fields together [Flurbereinigung (German), Melioration (Swiss), enclosure (English)] had to be pushed through by law in the face of resistance on the part of the peasants in many areas of the world. In contrast to the widely held opinion that the splitting up of the fields into small parcels was the expression of a—ultimately unsuccessful—law of inheritance, parcelling can now be understood as a risk strategy that was successful for a long period of time and was not rendered obsolete until the advent of modern agricultural methods.

Storekeeping and trade, combined with methods of resource management, provide the basis of the agrarian means of production whose origins date back some 10,000 years and which for some 5,000 years have reached a level of surplus production that has led peasants away from the older subsistence economies. According to our model, an impulse for the transition to agriculture resides in the attempt to make the flow of resources permanent by means of portfolio creation and, thus, in as close an approximation as possible to the average value. This had two fundamental consequences:

First, portfolio creation leads to an increase in the security of the food supply as well as to a rise in the level of the long-term flow of resources, as the lower edge of the fluctuations rises (cf. Figure 3). This means that the population can grow, as more people can be reliably fed in a certain area than before. Higher population density, however, leads to social conflicts and provides an incentive to develop institutional solutions for coping with them: The transition to complex societies is now evolutionarily favored. When storekeeping is organized and secured by state redistribution, highly stable complexes of urban civilizations can develop.

Second, when consumption consistently approaches the average level of resource flow, the luxury margin must shrink, having been produced by the discrepancy between the actual and the average consumption level. This means, however, that for the great majority of people, standards of living decline, while the amount of labor necessary to provide for the normal means of subsistence rises. Now the cultural pattern of "leisure orientation" loses its functional basis. In its place appears the maxim specific to agrarian societies that "in the sweat of thy face shalt thou eat bread."

In complex and socially stratified agrarian civilizations, ruling military, bureaucratic, landowning, or ecclesiastical classes tend to attract goods and labor as tribute, taxes, rents, or tithe for their peculiar purposes. This extraction of surplus may be regarded as wages for services rendered or as mere exploitation of the peasants. There is a major difficulty in the concept of surplus: When a tribute to a ruling class is paid, there may be a hidden reflux of services rendered, for example, military security, maintainance of infrastructure, peacekeeping, or providing the favors of supernatural powers. It is not easy to decide where merely a price is paid or a tribute extracted, that is, where the conflict between peasants and rulers resembles a conflict between seller and buyer about a price to be paid, or where it is a struggle between exploited and oppressing classes (Dalton 1974). In our view, "surplus" is a purely functional concept, meaning not an item that previously exists, to be appropriated by exploiters, but anything that may

leave the boundaries of a household, independent of any (material, infra-
structural, or symbolic) return. This definition, however, has the disadvan-
tage that it may not be able to differentiate between commodity exchange
and surplus attraction.

If surplus is to be connected with exploitation, at least the following conditions should be fulfilled (Boulding 1973):

1. The exploited classes suffer physical wants;
2. The ruling class lives in luxury (as related to the normal level of consumption);
3. The difference between both standards of living results from
   resource transfers from the exploited to the ruling classes; and
4. Physical wants of the exploited have their causes in inappropriate
   surplus expenditure by the ruling classes and could be redressed by
   more just forms of resource utilization.

However difficult it may be to define "surplus," some kind of social para-
sitism is a permanent feature of agrarian civilizations.

The normal state of agrarian civilizations thus essentially consists of a
combination of increased security of resource flow and a reduced level of
provisioning. Working long hours and hard, the majority of the population
lives in a state of constant material shortage, and luxury becomes the preserve of the upper classes. As wealth appears to be a result of labor, the ruling strata of agrarian societies tend to encourage population growth. So,
last but not least, pressure on natural resources and on the environment
increases.

Seen from this point of view, the meaning and function of luxury and
surplus assume a specific historical thrust. In subsistence economies,
which to a large extent refrain from technically controlling their resource
base, this "under-use" finds expression in a "leisure preference," which
indeed leads to a surplus of "work-free" time. The luxury of these societies
resides in an excess of time, which is simultaneously greatly prized in cultural terms.

The switch to an agricultural mode of production then triggered a development which likewise can be seen in formal terms as a strategy of risk
minimization but which has led to a completely novel economical and
social logic. In the agrarian civilizations, the values of productivity and
work enjoy high status, at least for the laboring lower classes, and luxury
presents itself almost exclusively in categories of commodities and goods.

Leisure preference is something solely for the free aristocrat—that is to say, for the aristocrat freed of the necessity of having to earn his living—but not for the working poor, for whom a curious work ethic now enters the world, which in modern Europe extended also to social classes that were actually in a position to dedicate their time to contemplation. At any rate, for the mass of the population in agrarian civilizations, there is no longer such a thing as "luxury," in terms of either time or commodities; for them, it remains banished to the Great Beyond of a land flowing with milk and honey where one only has to open one's mouth for the roast doves to fly into it.

THE RISK SPIRAL

The strategies of risk minimization by portfolio formation listed above already involve a price, which consists of the creation of specific new risks. From an early stage of history, therefore, the image of a risk spiral appears, whereby the reduction of a particular risk leads to new risk problems, which in their turn require new (risky) solutions.

Agriculture always produces a surplus capable of storage and transport, the use and distribution of which gives rise to conflicts that, directed at the outside world, can take on the form of conquest, plunder, and war and, directed at the society itself, can assume the shape of oppression and exploitation. The basic achievement of giving permanence to the flow of resources can thus not be relied upon. On the contrary, the killings, famines, and plagues brought about by conflict and now appearing on a massive scale usher in a new uncertainty factor, one which is not susceptible to elimination in the context of the agrarian civilizations.

Migrations, transport, and trade often led to conflict with competing groups or peoples. The difference between trade and robbery was frequently smudged. The distinction between merchant seafaring and piracy was also obscure. The battle for commodities, markets, and raw materials ultimately made war between states a fixed feature of the agrarian civilizations.

Storekeeping also led to novel risks. A supply of food is not merely capable of going off; it also tempts uninvited guests to partake of it and must, therefore, be protected against all manner of parasites—against insects, mice, and rats but also against human beings who might have a much greater incentive to plunder a food store than to laboriously seek nourishment themselves or even to till the fields. These constraints explain the development of the hygienic and politico-military conservation of the riches which are stored as provisions (on this complex, cf. McNeill 1980). The latter, above all, could take on seriously unpleasant aspects: The military caste of society deprived the peasantry of so many resources in the form of taxes and contributions that the difference between being plundered by alien bands of robbers and paying tribute to conquerors may not always have been extremely clear.

In combination with settling, greater security of provisions and a higher level of provisioning led to a relaxing of fertility controls, with the result that the populations of agrarian civilizations grew by comparison to hunter-gatherer societies by a factor of approximately a hundred. This, however, brought about the critical density of population that not only encouraged the spread of endemic diseases but also made genuine pandemics possible. For this reason, agrarian societies were afflicted much more than hunter-gatherer societies with infectious illnesses which, along with wars and famines, were the classic scourges of the agrarian civilizations.

The typical ambivalence of coping with risks and creating new ones revealed itself in various other areas in the agrarian civilizations. Money—that is, precious metals such as gold—can be stored for unlimited duration and is thus the ideal buffer, but it also attracts special parasites—thieves and tax collectors. The state creates infrastructures of trading communication, such as roads, ports, and canals, but it also produces artificial obstacles to mobility by setting up borders and levying customs duty. State redistribution of confiscated goods can alleviate crises when it is just and accentuate them when it is parasitic. It can thwart privately made provisions by means of severe taxation and then waste the resources thus gained. Finally, the market can act as a buffer against risks, but it also generates specific speculative risks and imbalances, above all for the specialists who are totally dependent on it.

TECHNOLOGICAL PROGRESS AND INDUSTRIALIZATION

A further factor involved in dealing with risk remains to be mentioned, however: actual technological progress. In our concept, it has two distinct functions. On the one hand, it helps lessen the effects of fluctuations, for instance, by facilitating irrigation in the event of drought. Its second and historically more important function, however, consists of raising the average value itself by increasing the productivity of a human habitat. This improves the security and simultaneously increases the level of provisioning, as long as population growth and environmental problems do not thwart this effect once more.

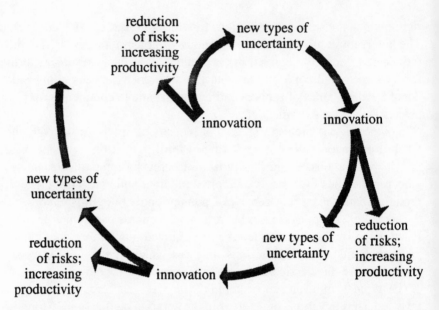

Figure 4. Risk spiral as a dynamizing principle in the development of complex societies: The reduction of a particular risk leads to new types of uncertainty which in their turn require further (risky) innovations. This mechanism creates the permanent innovation pressure responsible for the restless transformations in complex societies.

Technological progress was already an element of agricultural societies, especially in the stage of the agrarian civilizations, which always possessed an urban and commercial sector. This strategy began to take real effect only with the advent of industrialization, which was preceded by a phase of agricultural innovation. Now, for the first time, its most important result—that of raising the level of resource flow—became palpable for the mass of people in the industrialized countries, as the production figures grew faster than the population figures.

Here again, however, the effect of a risk spiral revealed itself. Manufacture essentially involves recourse to stocks of exhaustible resources (obsidian, copper, tin, iron), which already at an early stage created a compulsion to develop a permanent dynamic of prospecting and innovation. Here too, then, a specific dialectic of risk avoidance and risk creation came into being, introducing a curious dynamic in the sense of a continual "retreat in a forwards direction" (Wilkinson 1973).

It was only the late industrial society that brought about, on the basis of advanced technology, a novel type of mass affluence in complete contrast to the luxury type of preference for leisure. It is a question here of a fairytale superabundance of goods and energy, in which the population of the industrial zones luxuriates. Despite increased life expectancy, however, time remains in extremely short supply and is primarily conceived of as a means for the purpose of work and of acquiring riches. Unemployment, by contrast, is seen as a misfortune, and this not so much because of the low level of material affluence associated with it—a level nevertheless far above that of pre-industrial times—as because of the amount of useless time which the unemployed now have at their disposal and which they do not really know what to do with.

It may be, however, that this curious and paradoxical pattern is in historical terms merely a transitional phenomenon, the expression of a pioneer situation which will very soon be a thing of the past once more. In view of the growth of population across the globe, of shrinking resources, and of foreseeable tolerance limits in the natural world with regard to the effects of mass production and mass consumption, one can expect that for the majority of people, material consumption will drop back to a lower level once more, as was usual for the older civilizations. In the not-too-distant future, the era of mass affluence and widespread material luxury will be a thing of the past. It may be, however, that a new luxury dimension will then arise, possessed of a genuinely exclusive character: free use of space in the face of the overcrowding of the globe by human beings.

ACKNOWLEDGMENTS

The authors wish to thank Dr. Pierre Funck for his technical support.

NOTES

1. With an average covariance coefficient of 0.36, in the years 1335-1349, for instance—between the respective harvests, which constitutes a reasonable precondition for a palpable portfolio effect.
2. With a middle covariance coefficient of 0.6 even in the case of distances of 2-3 miles.

REFERENCES

Boulding, K. 1973. *The Economy of Love and Fear.* Belmont, CA: Wadsworth.
Caraco, T., S. Martindale, and T.S. Whittham. 1980. "An Empirical Demonstration of Risk-Sensitive Foraging Behaviour." *Animal Behaviour* 28: 820-830.
Cashdan, E. 1990. *Risk and Uncertainty in Tribal and Peasant Economies.* Boulder, CO: Westview Press.

Cohen, M.N. 1989. *Health and The Rise of Civilization.* New Haven, CT: Yale University Press.

Dalton, G. 1974. "How Exactly are Peasants Exploited?" *American Anthropologist* 76: 553-561.

Duerr, H.P. 1989. *Nacktheit und Scham. Der Mythos vom Zivilisationsprozeß* [Nakedness and Shame: The Myth of the Process of Civilisation], Vol. 1. Frankfurt am Main: Suhrkamp.

Groh, D. 1992. "Strategien, Zeit und Ressourcen. Risikominimierung, Unterproduktivität und Mußepräferenz - die zentralen Kategorien von Subsistenzökonomien" [Strategies, Time and Resources: Risk Minimisation, Underproductivity and Leisure Preference—The Central Categories of Subsistence Economies]. Pp. 54-113 in *Anthropologische Dimensionen der Geschichte* [Anthropological Dimensions of History], by D. Groh. Frankfurt am Main: Suhrkamp.

Harris, M. 1978. *Cannibals and Kings: The Origins of Culture.* New York: Vintage.

Harris, M., and E.B. Ross. 1987. *Death, Sex and Fertility: Population Regulation in Preindustrial and Developing Societies.* New York: Columbia University Press.

Hobbes, T. 1651. *Leviathan* I, 13.

Lee, R. 1968. "What Hunters Do For A Living, Or, How to Make Out on Scarce Resources." Pp. 30-43 in *Man the Hunter*, edited by R. Lee and I DeVore. Chicago: Aldine.

_____. 1979. *The !Kung San.* Cambridge: Cambridge University Press.

Lovejoy, A.J., and G. Boas. 1935. *Primitivism and Related Ideas in Antiquity.* Baltimore, MD: Johns Hopkins Press.

Mace, R. 1993. "Nomadic Pastoralists Adopt Subsistence Strategies that Maximise Long-term Household Survival." *Behavioural Ecology and Sociobiology* 33: 329-334.

McCloskey, D.L. 1976. "English Open Fields as Behavior Towards Risks." Pp. 124-170 in *Research in Economic History*, Vol. 1, edited by P. Uselding. Greenwich, CT: JAI Press.

McNeill, W.H. 1980. *The Human Condition: An Ecological and Historical View.* Princeton, NJ: Princeton University Press.

Mead, M. 1928. *Coming of Age in Samoa: A Psychological Study of Primitive Youth for Western Civilization.* New York: W. Morrow.

Morgan, L.H. 1877. *Ancient Society.* New York: Holt.

Sahlins, M. 1974. *Stone Age Economics.* London: Tavistock.

Slobodkin, L.B., and A. Rapoport 1974. "An Optimal Strategy of Evolution." *The Quarterly Review of Biology* 49: 181-200 .

Wilkinson, R.G. 1973. *Poverty and Progress.* London: Methuen.

ACCUMULATION, DEFORESTATION, AND WORLD ECOLOGICAL DEGRADATION, 2500 B.C. TO A.D. 1990

Sing C. Chew

ABSTRACT

World history for at least the last 5,000 years has been underscored by the ceaseless accumulation of capital and punctuated by long cycles of economic growth and stagnation. Such economic processes occur via trade and production processes within the context of competitive rivalries among empires, civilizations, and states, leading to core-periphery (powerful-weak) relations that further structure socioeconomic exchange and world trade. To date, what is missing in understanding world historical transformations is the nature-societal relation. This is as primary as the economic relation in the self-expansionary processes of societal systems.

I suggest that ecological limits also become the limits of the socioeconomic processes of empires, civilizations, and nation-states, and the interplay between ecological limits and the dynamics of societal systems defines

Advances in Human Ecology, Volume 6, pages 221-255.
Copyright © 1997 by JAI Press Inc.
All rights of reproduction in any form reserved.
ISBN: 0-7623-0257-7

the historical tendencies and expansionary trajectories of the human enter-prise. To support this contention, I offer a closer examination at how the historical process of accumulation of wealth, core-periphery relations, and cycles of expansion and stagnation are related and conditioned by societal relations with nature over world history. In this context, I focus on the production, circulation, and consumption of timber and wood products, including the subsequent outcomes of such activities (such as deforestation and ecological degradation), to amplify historically this nature-societal relational dynamic.

INTRODUCTION

Over world history, the relationship of human societies with nature has been transformative. In most cases, this nature-societal relation has been exploitative of nature, engendered primarily to meet the materialistic requirements of hierarchical systems of social organizations. Such practices have been repeated over five thousand years of world history from the ancient civilizations of Mesopotamia, China, Mesoamerica, and India to the current era (Chew 1995; Perlin 1989; Thomas 1983). Viewed from this long-term perspective, our relationship with nature has not changed significantly.

Historically, we find that to maintain an economic surplus there must be continual exploitation of nature to meet certain consumptive levels in societal production and reproduction (Ponting 1991; Frank 1993). These socioeconomic activities have often resulted in class struggles, wars, and in some cases, genocide. William McNeill (1992) has defined such exploitative socioeconomic relations over world history as macroparasitic. This macroparasitism also engenders ecological degradation resulting in loss of species diversity, polluted oceans and rivers, siltation, population losses as a consequence of flooding, and health issues—the most extreme scenario being the collapse of civilizations such as the Harrapan, the Mayan, and the Mesopotamian (Perlin 1989; Ponting 1991; Chew 1995). Therefore, McNeill's characterization can also be extended to the nature-societal dimension.

Among some scholars in the social sciences, there is an increasing interest in understanding world transformations as a consequence of a set of socioeconomic processes that have been occurring over the very long term within the context of a world system or systems (Frank 1991, 1993; Chase-Dunn and Hall 1996; Ekholm and Friedman 1982; Modelski and Thompson 1996; Chew 1995; Wilkinson 1981). Contrary to Immanuel Wallerstein's view (1974, 1991, 1992), Andre Gunder Frank (1993) has suggested that world history for at least the last five thousand years has been underscored by the ceaseless accumulation of capital and punctuated by long cycles of economic growth and stagnation. Such economic processes occur via trade and production processes within the context of competitive rivalries between empires, civilizations, and states leading to core-periphery (powerful-weak) relations that further structure socioeconomic exchange and world trade. To date, what is missing from such an approach to understanding world historical transformations is the nature-societal relation. The studies to date have overwhelmingly focused on the human-human side of the equation, that is, on the social (political-economic) relations of classes, regions, and civilizations, empires, and states. But are these supposedly materialist social factors and conditions sufficient to account for transformations in the *longue duree*? To be minimally materialist, the basis of societal reproduction (in a broader context) must be viewed also through our relations with nature. It is underscored in the early civilizations of Egypt, Mesopotamia, Indus, and the Hwang Ho, whose social reproductions were contingent on economic (production of surplus) and ecological relations. These societies' reproductive and expansionary capacities were conditioned by their specific ecological surroundings coupled with the needed search-exchange in other ecological landscapes for the natural resources (timber, metals, and certain stone) they lacked or had already exhausted by unsustainable exploitation and accumulation. *Thus, the ecological relation is as primary as the economic relation in the self-expansionary processes of these societal systems.*

Viewed in this manner, the societal and ecological worlds interact in a fashion whereby nature's rhythms impact on the dynamics of social and economic life and vice versa. For example, changes in climatological trends, such as temperature, impact crop harvests which, in turn, impact grain prices or the migration of people (see, e.g., Ladurie 1971). *Pari passu*, societal systems, with their productive and consumptive life styles, generate degradative effects on nature (destruction of species, global warming, and so forth) which, in turn, loop back to impact the dynamics of social and economic life (such as flooding, crop failures, port siltation, and so forth). Thus, the ceaseless accumulation of wealth over world history seems ultimately self-defeating in that nature as the underlying basis of the accumulation equation provides, conditions, and inhibits this process and, thus, establishes its limits. It suggests that *ecological limits become also the limits of socioeconomic processes of empires, civilizations, and nation-*

states, and the interplay between ecological limits and the dynamics of societal systems defines the historical tendencies and expansionary trajectories of the human enterprise.

Does world history reflect the tenor of the above delineations of the nature-societal dimension? Does it affirm our claim that we should not forsake the nature-societal nexus and that this nexus is intertwined with socioeconomic relations in the reproduction of societal systems? To support this contention, I offer a closer examination of how the historical process of accumulation of wealth, core-periphery relations, and cycles of expansion and stagnation are related and conditioned by the societal relations with nature over world history. In this context, I focus on the production, circulation, and consumption of timber and wood products, including the subsequent outcomes of such activities (such as deforestation and ecological degradation) to amplify historically this nature-societal relational dynamic.

NATURE-SOCIETAL RELATIONS, 2500 B.C. TO A.D. 1990

If one examines world history over the long term, one can see that it has involved the extraction of natural resources such as wood to meet the reproductive and expansionary dynamics of societies and civilizations. Wood has been used for fuel, building materials, and other production needs for at least over 5,000 years of world history (Perlin 1989; Ponting 1991). As such, it has facilitated directly or indirectly the process of accumulation in the world system. The long history of wood utilization to meet human material needs has been exploitative (macroparasitic) in nature, and has often been undertaken within and between trade and territorial (and including extra-territorial) linkages of societies, kingdoms, civilizations, and empires, beginning from the ancient civilizations of Mesopotamia, China, and India and extending to the present (Perlin 1989; Attenborough 1987; Butzer 1971; Marsh 1965; Thirgood 1981). Over world history, those states or kingdoms that were located hierarchically in more advantageous positions in the total accumulation process could extract wood products to meet their social and economic transformative needs through tributes, conquests, or trading relations. Notwithstanding the ecological damages, this socioeconomic linkage engenders center-periphery relations and, over time, structures the dynamics of the trading patterns as well as the political economy of the region in question.

The utilization of wood in human history has occurred in two main ways. The first was for basic materials for buildings, ships, mine shafts, and storage purposes. The other manner of utilization was in the production of commodities for exchange (such as pottery, beads, and so forth), as fuel for domestic use (in the form of charcoal) and for the further extraction of other resources for production processes (such as the making of bronze and iron), and as luxuries such as scented wood products. Wood, therefore, has been a basic commodity underlining the reproductive aspects of kingdoms, states, and civilizations. As such, it facilitates the accumulation of capital or, depending upon cultural needs and availability, is itself a precious commodity for exchange in the overall surplus generation.

2500 B.C. to 500 B.C.:   Mesopotamia, India, Crete, and Mycenaean Greece

One may see a continuity in this pattern of accumulation not only in recent history but even as far back as early China, third millennium B.C. Mesopotamia, Bronze Age Crete, Mycenaean Greece, Greece of the classical period, second century B.C. Rome, and India as early as 2000 B.C. Throughout these civilizations, we find the utilization of wood and the search for wood to reproduce the needs of the kingdoms, states, and civilizations. If we examine world history from Mesopotamia to the Indus valley around 2500 B.C., we find the sustained utilization of wood and wood products to meet various materialistic, consumptive, and productive needs. In third millennium B.C., the kingdoms of Lagash and Ur utilized wood (for example, cedar) for buildings (including temples), ship construction, and canals. The consumptive needs were high, as Lagash and Ur had populations of 37,000 and 65,000, respectively. The widespread use of tools requiring wooden handles and the need for wooden household furniture and utensils further increased wood imports. With the increased use of bronze, wood was required as fuel for foundries. As a core center of accumulation, besides causing the exploitation of local forests, imports were exchanged from the Ammanus mountains (southwest Turkey), southeast Arabia, and as far away as the Indus region of India (Tibbetts 1956; Kohl 1987; Edens 1990; Ratnagar 1981). Wood was in such high demand that during periods of accelerated economic expansion, its value was equivalent to precious stones, and some types of wood were stored in the royal treasury (Perlin 1989, p. 41). Expeditions were sent to seek new sources when wood supply was constricted, and Eastern Turkey and the island of Crete had replaced India as a supply source by the time of the reign of Hammurabi. In this context, luxury goods from Babylon were traded for

Cretan wood. What we observe is the overall expansion of socioeconomic growth sustained through wood consumption in production activities, and the exchange of manufactured products such as textiles for wood imports. Within a world-systems analysis context, we are also observing core-peripheral relations—Mesopotamia during this period was a center of accumulation, exchanging with lesser-"developed" areas for commodities to enhance its wealth.

In other parts of the world system, such as South Asia and the Far East, the first assaults on the forest cover came in India as early as the third millennium B.C., in northern China around the Hwang Ho river basin from about the same period, and in Southeast Asia (such as the Malayan peninsula) about 2500 B.C. (Wheatley 1961; Tibbetts 1956). Evidence of the volume of wood extracted is lacking except for qualitative comments (Wheatley 1961; Dunn 1975; Gadgil and Guha 1993; Ponting 1991). We do know also of trade exchange between India, Mesopotamia, and the Mediterranean zone as far back as the third millennium B.C. Tibetts (1956, pp. 183-184) reported evidence of Indian wood found in early Sumerian cities and on inscriptions indicating the importation of wood from India as early as 2000 B.C. In terms of trade between Mesopotamia and China, Kennedy (1898, as cited by Tibbetts 1956) has suggested that trade began as early as the seventh century B.C.

In the Indus valley around mid-2500 B.C., the Harappan civilization flourished through its trade contacts with Mesopotamia. Wood, stone, metals, cereals, oils, and other items were exchanged with Mesopotamia from cities such as Meluhha. Timber and wood products were part of the trade exchange as the Indus valley was richly forested, with a plentiful supply of wildlife during this period (Ponting 1991). Teak, fir, and pine were extracted from the Western Ghats, the Jammu ranges, and the Panjab piedmont. Widescale building of temples and palaces was the order of the day, requiring mud bricks that had to be manufactured by drying them in ovens fuelled by wood (Wheeler 1968; Marshall 1931). Along with this massive utilization of wood, and including trade exchanges in timber with Mesopotamia, there was widescale deforestation of the surrounding areas for precious metals, such as gold in the mountains near Meluhha. In addition, the utilization of copper and bronze for farming implements and other household items also required wood as a fuel source for their manufacture, although cow dung was also used.

The utilization of wood proceeded in Bronze Age Crete. With its trade in wood products with the near East, including Mari and Babylon, Crete emerged as a center of accumulation. Growth was concentrated at such places as Knossos (population 30,000 at 1360 B.C.) where an abundant supply of timber fuelled the transformations (Chandler 1974, p. 79). As in Mesopotamia, massive utilization of wood was required for shipbuilding, the manufacture of bronze and pottery, and building construction, including palaces and administrative offices. At the height of Minoan power, there was extensive demand for wood to build merchant ships and warships as a consequence of the increased trade between Crete and mainland Mycenaen Greece and the eastern Mediterranean (Meiggs 1982, p. 97).

The trade between Crete and Mycenaen Greece was facilitated by the bountiful forests on the Greek mainland. This control of abundant wood resources allowed the Mycenaens to demand a hefty sum for the wood that the Cretans needed badly for their continued socioeconomic growth. As a consequence, a transfer of wealth occurred from Crete to the mainland (Perlin 1989. p. 54). As a center of accumulation in the Mediterranean, Mycenae by 1350 B.C. had a similarly sized population as Knossos and was in an expansionary phase, building palaces and manufacturing bronze products and pottery. There were also extensive trade relations with southern Italy and the Syrian-Palestinian coastal areas (Chandler 1974, p. 79; Perlin 1989).

Between the sixth and the fourth centuries B.C., with the decline of Crete as a regional power, widespread maritime trade took place between mainland Greece and the surrounding areas of Asia Minor. Large merchant fleets were built and colonies were established for the purposes of trade and "to ensure a flow of essential commodities and materials to the mother city" (Thirgood 1981, p. 9). Colonization also spread, with settlements in southern Italy and Sicily to the west and around the Black Sea, as well as in northern Greece (Meiggs 1982, p. 121). These settlements became centers of commerce and manufacturing activities in addition to being agricultural centers drawing on their hinterlands for supplies. All in all, this socioeconomic expansion facilitated the accumulation processes of the city-states of mainland Greece. In terms of timber resources, the trade extended to cover the central and eastern Mediterranean, including the Black Sea, Asia Minor, and the Caucasus (Thirgood 1981). There was also exchange of special wood products such as teak and ebony with India, paid for by manufactured goods, thus underscoring the continuity of trading relations with India, seen as far back as the third millennium B.C. with Mesopotamia. With continued economic expansion, increased urbanization followed, leading to further pressure for increased agricultural production that required the colonies in the outlying areas to provide agricultural produce for the urbanized environments. Industrial products were manufactured for exchange with the basic primary resources located in the hinterland, thus requiring further deforestation. Hinterland areas such as eastern Greece and western Asia Minor were subsequently depleted of forests. Other coastal regions of Phoenicia, Syria, Egypt, and Italy provided the grain and food imports for the needs of an urbanized Hellenistic Greece. Of course, deforestation occurred in these areas as well.

This expansionary period generated an increase in the production of manufactured commodities requiring basic natural resources, such as iron ore, for production processes. The hinterland areas of the Mediterranean, such as Asia Minor, Phoenicia, Syria, Sardinia, Elba, and Spain, supplied these natural resources. Where iron was exported to mainland Greece, forests were depleted for metal smelting. A million acres of productive woodland was required to meet the needs of a single metallurgical center during the classical age (Thirgood 1989, p. 56). Furthermore, with developments in the banking system from the sixth century B.C. onwards, profitable investments were made in agricultural production. The rising prices for agricultural produce between the sixth and fourth centuries B.C., and the increasing use of manures coupled with controlled times for plowing and harvesting, made farming more profitable and thus facilitated this expansionary process. No doubt, such financial investments further spurred on deforestation.

Ecological Degradation of the Period

The extraction of wood resources most often led to ecological degradation, and in some cases to the collapse of empires and civilizations (Perlin 1989; Ponting 1991). Thus, for ancient empires and hegemonic powers, the deforestation occurred not only within the proximity of the developed centers but also in the outer domains of these empires or in their external arenas. The ceaseless assault on forests for fuel for the accumulation process had hitherto known no bounds. The process continually shifted to other areas that were more fertile and less inhibitive (that is, areas which offered the availability of trees and a lower level of social resistance) for continued accumulation after nearby forest resources had been exhausted (Chew 1992, 1995). Such widescale deforestation to meet the materialistic needs of societies, states, kingdoms, and civilizations is by no means the only avenue by which deforestation has occurred over world history. The clearing of land for agriculture, so that grain and livestock needs could be met for local consumption or for core centers of accumulation, has to be underscored. Notably, it has been stated that the advent of agriculture to meet human consumption and exchange needs was one of the main causes of deforestation and environmental degradation in early world history (Ponting 1991). The outcomes of these ecological degradations over world history resulted in flooding, loss of topsoil, temperature increases, biodiversity losses, and so forth (Roberts 1989). In fact, they have recurred throughout world history for the last 5,000 years.

We can observe such themes in early civilizations, where forests were intensively utilized to meet socioeconomic needs. These early social systems were dependent on the production of agricultural surplus to reproduce social hierarchies which had an increasingly high number of priests, rulers, bureaucrats, and soldiers. This entailed land clearing for agricultural production, thus leading to deforestation, which then impacted future crop yields. For example, in Mesopotamia, deforestation created soil erosion, leading to the siltation of the rivers and the waterlogging of the land. As the land became waterlogged, the water table was raised, causing more mineral salts to be brought to the surface, where the high summer temperatures (40°C) produced a thick layer of salt on the land, thus lowering crop yields (Ponting 1991; Jacobsen and Adams 1958). Perlin (1989, p. 43) has indicated that excessive deforestation of the northern mountains of Mesopotamia to support the socioeconomic development of the south and to maintain a standing army around 2400 B.C. led to an accumulation of mineral salts in the irrigated farmlands of southern Mesopotamia, which over the course of 300 years led to declining crop yields of 42 percent. By 1800 B.C., crop yields were only about a third of the Early Dynastic period, and no wheat was grown in southern Mesopotamia (Ponting 1991, p. 72). Thus, when agricultural production went down in Sumeria due to increasing salinity, "the superstructure of administrators, traders, artisans, warriors, and priests that comprised this civilization could not survive" (Perlin 1989, p. 43). In this way, deforestation can be seen as one of the factors leading to the demise of this center of accumulation in Mesopotamia, which shifted to Babylonia around 1700 B.C. The declining crop yields meant that surplus could not be generated as effectively, which increased the difficulties of maintaining a large bureaucracy and army, which in the long run made the state vulnerable to conquest. By 1800 B.C., Sumer had declined in significance and was underpopulated. After a while, second millennium B.C.

Babylon also suffered the same condition whereby the scarcity of wood led to increasing fuel costs and increasing prices for wooden articles. A parallel to these kinds of conditions is the Indus Valley civilization. As in Mesopotamia, its complex, hierarchical social system flooded the Indus by means of extensive works and increased agricultural production to feed the ruling elite, the priests, and the warriors. Deforestation was the order of the day, as trees were even cut down for fuel to dry bricks for buildings and palaces. Along with deforestation, salinization also occurred in the agricultural areas, leading to the further inability of the social system that supported its hierarchical system (Ponting 1991; Hoffman 1980, p. 34). At about 1900 B.C., the Indus civilization came to an end, with environmental degradation as a contributing factor in its decline (Ponting 1991, p. 73).

In the late Bronze Age, these conditions were repeated again in Mycenaean Greece, where the deforestation of the hillsides resulted in large amounts of earth and water draining from the slopes onto the Plain of Argos, filling up the streams and leading to extensive flooding. Tiryns and its agricultural lands were affected by flood waters. Pylos' harbor was also impacted in terms of siltation, as was the island of Melos. With the loss of topsoil, agricultural production in areas such as Messinia automatically diminished.

Decreasing crop yields strained the reproducibility of the society. In addition, the centers of accumulation of Mesopotamia, such as Lagash and Ur, exploited their peripheral hinterlands, from eastern Turkey to even as far as Lebanon, for wood. Severe environmental degradation occurred, resulting in depopulation.

Extensive deforestation also impacted the production processes of these social systems. Wood scarcity at Knossos on the island of Crete forced changes in production locations or resulted in their closures. Over time, the continued reliance on imported wood from Mycenae and Pylos resulted in the transfer of wealth from Crete to the Greek mainland (Perlin 1989, p. 54). Mycenae, though benefiting from its abundant supply of forested areas, did not pursue a sustainable yield for its forests. By the late Bronze Age, where pasture land was cleared in Mycenae for sheep grazing, metallurgical production works had to be relocated to lesser populated areas where wood supplies were more available. Mycenaean prosperity built on metallurgical and pottery works suffered with the decline in fuel supplies, especially the pottery works at Berbati and Zygories. Population migration followed the closure of these manufacturing centers, and the abandonment of Phylakopi coincided with the deforestation of Melos, where the town was located. The same situation occurred at Berbati, Midea, Prosymna, and Zygories in 1200 B.C., and there were also population losses in Mycenae, Pylos, and Tiryns. Throughout Mycenaean times, towns and settlements disappeared; for example, in southwest Peloponnese, the number dropped from 150 to 14. Other regions experienced similar declines, with Laconias, Argolid, Corinthia, Attica, Boetia, Phocis, and Locris registering population losses. Overall, by the eleventh century B.C., the number of inhabitants in these regions had fallen by 75 percent (Perlin 1989).

400 B.C. to 500 A.D.: Classical Greece and Rome

By the beginning of the fifth century B.C., mainland Greece was on a socioeconomic growth trajectory, especially the city-state of Athens. Becoming a center of accumulation required a strong navy to ensure control of the trading route and to thwart aggression from Persia. Because the Persians controlled much of northern Greece, Athens initially had to rely on local timber for its shipbuilding program. By 357 B.C., it had an inventory of 285 triremes (Meiggs 1982, p. 123). With the defeat of the Persians in 469 B.C., Athens' position as a center of accumulation in the Mediterranean was ensured, and socioeconomic expansion ensued. This resulted in rapid urbanization, requiring large quantities of wood for buildings and houses. By this time, the city's population had grown to nearly 200,000 (Chandler 1974, p. 79). This demographic surge, like the previous period in world history, further increased the demand for wood for fuel (charcoal) and for the manufacture of commodities (Perlin 1989, p. 86). As a consequence, the price of wood rose. This prompted the Athenians to search for other wood sources through conquest and colonization. One such source was Amphipolis, which, in 495 B.C., Athens colonized to exploit its forests. The extensive use of timber for shipbuilding to fight the Peloponnesian War had led Athens to cut its own surrounding forests and to rely on Amphipolis (Meiggs 1982; Perlin 1989). When the latter source was cut off, Athens then relied on Macedonia (Meiggs 1982). Following its defeat, the surrounding forests of Athens were devastated by years of forest exploitation.

Like Hellenistic Greece, Rome experienced an increased pace of socioeconomic activity when it became a center of accumulation. Urbanization followed, with the city expanding over its surrounding hillsides. This promoted deforestation. To meet its timber needs, Rome subjugated its surrounding hinterland, such as Liguria and Umbria. It conquered Etruria in the second century B.C. to provide timber for its shipbuilding needs (Perlin

1989). In addition, the forests of the Po valley were exploited. As the power of Rome grew, expansion followed into western Europe and North Africa. There was pressure to increase the food supply, and many of the outer areas of the empire were transformed into granaries to feed the population of Italy, particularly after 58 B.C., when Roman citizens received free grain (Ponting 1991; Thirgood 1989). The forested areas in North Africa were cut and cultivated, and by the first century A.D., the Roman province of Africa each year sent enough grain to feed a million people for two-thirds of a year (Meiggs 1982, p. 374). Besides grain, other natural resources were sought to fuel the processes of accumulation. Iberia was conquered for its silver, and copper was sought in Cyprus. Mining for these resources meant further deforestation. By the second century A.D., these places had become the manufacturing and resource bases for the empire. With this level of economic activity, an extensive regional trade ensued and cheap manufactured products were produced (Thirgood 1989, p. 29).

Ecological Degradation of the Period

As with the earlier period from 2500 B.C. to 500 B.C., we continue to observe degradative outcomes in second-century A.D. Rome and throughout the Roman empire. Degradation continued into the fourth century A.D., especially in the outlying provinces where trees were cut for mining, manufacturing, and agriculture (Shaw 1981, p. 392; Ponting 1991, p. 77). For example, during the four hundred years of silver smelting in Iberia, used to meet the growth of Rome, 500 million trees were cut (Perlin 1989, p. 125). North Africa's forests were devastated to make way for grain cultivation. Morocco lost 12.5 million acres of forests during the Roman period. Attenborough (1987, p. 118), writing on the excesses of the Romans during the third century A.D., revealed the level of deforestation that occurred starting from the Mediterranean areas northward into Europe. Randsborg (1991) has also noted that this deforestation was a result of agricultural expansion. According to Attenborough (1987, p. 117), the Romans' view of the natural world—much similar to the predominant anthropocentric viewpoint of the late twentieth century—was of a world to be ravished and plundered. In the words of Cicero:

> We are the absolute masters of what the earth produces. We enjoy the mountains and the plains, the rivers are ours. We sow the seed and plant the trees. We fertilize the earth...we stop, direct, and turn the rivers, in short by our hands we endeavor, by our various operations in this world, to make, as it were another nature (Thirgood 1989, pp. 29-30).

As a consequence, lands which were high yielding in terms of grain production, such as North Africa between Egypt and Morocco, over time became barren. Other Roman writers, such as Pliny and Strabo, indicated that deforestation in Tuscany and Campania was intended to supply the needs of Rome (Shaw 1981; Perlin 1989; Thirgood 1989).

With deforestation we also find soil erosion, which meant siltation of ports and low-lying areas. This occurred in classical Greece and also in the Roman empire. The port of Paestum in southern Italy silted up, and the town of Ravenna lost its access to the sea (Ponting 1991). Ostia, the port for Rome, managed to survive after major dock reconstruction. Because of siltation, the cities of Greek Asia Minor, such as Priene, Myus, and Ephesus (fifth century B.C. to second century A.D.) over time became inland cities, where once they had been located on the coastline of Asia Minor facing the Aegean Sea.

Deforestation also can force the relocation of industries, which over time may lead to loss of commercial and productive dominance. By the fourth century B.C., Athens experienced this fate when it faced shortages of wood and thus had to relocate its metal industries closer to sources of wood. Denied further access to other forestland in Greece, Athens continued to face shortages, and its power on the Greek mainland slowly declined (Perlin 1989, p. 100). Toward the last years of Rome's power, industries had to be relocated to areas in Europe so that fuel sources would be closer to production. By the second half of the first century A.D., ceramic factories were established in southern France and, in turn, their products were re-exported back to Rome. Glass and iron manufacturing were also exported to regions such as Gaul and southern France. With the level of deforestation, the various bronze, glass, and pottery manufacturing processes established in southern France had to curtail their production levels. The iron mines in Britain and the copper mines in Cyprus also suffered due to the lack of adequate wood supplies.

500 A.D. to 1500 A.D.:   Asia, the Mediterranean, and Europe

Trade continued to flourish between India, the Far East, and the Mediterranean in the first century A.D. (Tibetts 1956; Colless 1969; Abu-Lughod 1989). For Asia, the core centers of accumulation were located in China, South Asia, and the city-states of Southeast Asia. Records reflect the trade in wood products between China and Southeast Asia (including Japan), with the former supplying manufactured products and the latter natural resources, of which wood was one of the primary commodities (Wheatley 1959, 1961; Wang 1958; Wolters 1967; Coedes 1983; Yamamoto 1981; Totman 1989). The height of economic expansion of the world system was during the period of the Tang and Sung Dynasties, where city-states located in the Malay archipelago, Annam, Java, and Sumatra had trade exchanges (including wood products) with China. These trading relations continued into the fifteenth century prior to the arrival of the Europeans. Wheatley (1959, p. 19) has noted that Chinese trade envoys were sent by the Han emperor Wu, during his reign from 141-87 B.C., to explore the South seas as far as the Bay of Bengal. Besides luxuries and spices, brazil wood, cotton cloth, swords, sandal wood, camphor, rugs, and even African slaves were traded (Wheatley 1959; Lim 1992; Lian 1988). The lines of commerce were quite well established before the first century A.D., and timber and other wood products were part of this trade exchange. According to Wang (1958, p. 21), there was trade between China and ports on the Indian Ocean by the second half of the first century B.C.

With the unification of China by 221 B.C., Chinese expansion to the south was pursued. The classic trade exchange between China as a core center of accumulation and the kingdoms and city-states in Southeast Asia—economically less transformed but not without some substantial economic strengths—can be seen in the type of commodities that were traded. China supplied silk and manufactured commodities such as porcelain and metal products, while Southeast Asia supplied natural resource products and spices. Records reflect this trade, including wood products after the first century A.D. By the second century A.D., the total number of people residing in the Tonking delta came to about 1,372,290 (Wang 1958, p. 18). Tribute missions from the city-states and kingdoms of Southeast Asia, and also from south India, started arriving in China during the second century A.D. Such missions, according to Wang (1958, p. 119), included the payment of tribute with wood products, so that political and economic concessions could be obtained.[1] For example, the state of Lin-yi in 433 A.D. provided tribute to obtain territorial concessions in Chiao-chou, and the state of Funan in 484 A.D. demanded relief from the incursions of Lin-yi. In addition to wood, precious metals such as gold, silver, and copper were also provided to the Chinese court. A mission from Lin-yi brought tribute of 10,000 kati of gold, 100,000 kati of silver, and 300,000 kati of copper (Wang 1958, p. 52). Such tribute missions increased in volume over time; by the Tang dynasty, a total of 64 missions were recorded (Wang 1958, pp. 122-123). By the third century A.D., the trade in wood products had grown. Gharu wood was imported to southern China, which involved merchants from the Malay archipelago, Sumatra, and even as far as Ceylon. City-states such as Lo-yueh (near Hanoi) were the collection centers (semiperipheral states?) for forest products, while P'eng-feng produced lakawood (Wheatley 1961, pp. 60, 71). Farther south, Tun-sun situated on the Malay peninsula was a dependency of the state of Fu-nan in Indochina.[2] Judging from the amount of tribute provided to China, these states must have been prosperous and economically developed. Lo-yueh, for example, was said to have palaces and 20,000 soldiers. By the late sixth century A.D., the kingdom of Ch'ih-t'u (situated on the Isthmus of Kra) became a substantial power in the region with the fall of Fu-nan (Wheatley 1961; Wang 1958; Dunn 1975).

The trading relationships between the kingdoms and city-states of Southeast Asia were buttressed further with the arrival of the Persian and Arab merchants by the seventh century A.D., found mainly in Canton, Hanoi, Yang-chou, and Ch'uan-chou.[3] Wheatley (1959, p. 28) has noted that these were mainly Middle Eastern Muslims, who established their own mosques and suks in these areas and were governed by their own sheikhs. The power and number of these merchants grew; by the mid-eighth century A.D., they were of such substantial strength that, for example, they were able to settle their disagreements with the Chinese by burning buildings and godowns (warehouses) in Canton in 758 A.D. Along with the Arab and Persian, the Srivijaya kingdom, situated in Sumatra, was developing into a regional power in Southeast Asia (Wolters 1967, 1982; Wang 1958; Wheatley 1961; Coedes 1983). Srivijaya maintained its commercial position in Southeast Asia for at least two centuries (8th-9th).

By the end of the ninth century A.D., Southeast Asia was known to a large number of Arab, Indian, and Persian traders, and the goods from this area were transported back to the Middle East (Tibbetts 1957; Colless 1969; Dunn 1975). With the sea route paralleling the Silk route of Central Asia to the Mediterranean and the West, the zone circumscribing the Arabian Sea, the Bay of Bengal, the Straits of Malacca, and the South China Sea was one of trade and exchanges with cities and kingdoms located in southern Arabia, southern India, the Malayan archipelago, Sumatra, Java, Indochina, and southern China, as well as with the Mediterranean (Wang 1958; Wheatley 1961).[4] In view of the trade interconnections and economic expansion and growth, this network could be considered an (Asian?) world system, with the core centers of accumulation as listed above. With the prevailing monsoon winds that blow in various directions during different times of the year, the merchant ships could sail to different ports depending on the season (Chaudhuri 1985; Wang 1958; Wheatley 1961). The Northeast and Southwest Monsoons would bring one set of merchants from one part of this zone to Southeast Asia, and at the same time take another set of merchants home. For example, when the Arab and Indian shipping reached Southeast Asia on the Southwest Monsoon, the Chinese junks would be sailing home. It seems, therefore, that Southeast Asia was the nexus of the system and that it continued to play a connecting role up to the coming of the Portuguese in the fifteenth century.[5] The volume of trade exchange, which included wood products, continued, and by the middle of the Tang Dynasty, the ships of Ceylon—over 200 feet long and carrying 600 to 700 persons—were plying the waters of the South China Sea. These ships were probably built in China because of the abundant timber resources in the coastal areas of Chang-chou, and in Ch'ao-chou, Hsun-chou, Lui-chou, and Chin-chou of Kwangtung province (Wheatley 1958, p. 109). By 987 A.D., during the Sung Dynasty, the southern maritime trade provided a fifth of the total cash revenue of the state; such a high volume led the state to support overseas missions to induce foreign traders to come trade at Chinese ports.[6] With its overland Central Asian trade routes cut off after 1127 A.D., China proceeded to exploit the sea route of the South China Sea. Ebony, gharu wood, laka wood, pandan matting, cardamons, ivory, rhinocerous horns, and beeswax were imported from Asia and India (Dunn 1975; Wheatley 1959). To illustrate the increase in trade between 1049 to 1053, the annual import of tusks, rhinocerous horns, pearls, aromatics and incense was about 53,000 units; after 1175 A.D. they reached 500,000 units. The increase in activity naturally led to the emergence of a powerful merchant group, which gradually came to manage all the major governmental monopolies of this trade. With the Mongol control of south China in 1277, trade with the rest of Asia and the Middle East was further encouraged. Southern China by the end of the thirteenth century had about 85 to 90 percent of the country's population. The region experienced an expansionary phase from the ninth to the fourteenth centuries, during which industry intensified and agricultural production increased. This must have led to deforestation to clear land for farming and to provide wood to fuel the kilns for the manufacture of pottery and for metals. Metallic currency was used as payment for products from Asia and the Middle East and, along with this, silk and pottery were exported. According to Wheatley (1959) and Yamamoto (1981), by the time of the Sung dynasty, there was a deficit in the Nanhai Trade which was covered with payments in bullion and coins. During this period, China also experienced a metallic coin shortage (Yamamoto 1981, p. 24).

After the fall of the Yuan dynasty, the Ming dynasty proceeded to build the Nanhai Trade. China had a sizable navy; by the end of the fourteenth century, it had 3500 ships of which over 1700 were warships and 400 were armed transports (Lo 1958). From the type of wood imported, we can surmise that China utilized its own pine and cedar forests for shipbuilding, which could be obtained from the coastal areas of southern China. Between 1403 to 1433, seven expeditions were sent, each comprising as many as 62 vessels carrying 37,000 soldiers (Yamamoto 1981). Under the command of Admiral Cheng Ho, these expeditions sailed as far as Mecca, Ormuz, Aden, Mogadishu, and Juda. When these expeditions ended in 1433, the Chinese fleet was withdrawn from the Indian Ocean leaving a power vacuum which was filled by the Portuguese later in the century.[7]

The pattern of accumulation is also repeated in Japan. Forests were cleared from 200 A.D. onwards for fuel as charcoal to produce metal implements, for domestic use, and also for building fortresses, temples, shrines and the cultivation of crops. For Totman (1989, p. 10), "agriculture and metallurgy were the human innovations that most dramatically affected prehistoric Japanese forests." Forest exploitation was continuous from 600 A.D. to 1670 A.D. From 600 A.D. to 850 A.D. onwards, within an economic expansionary environment, the ruling elites of Japan proceeded with a building boom, constructing palaces, mansions, shrines, temples, and monasteries. These buildings were erected near the capital cities of Nara and Heian (Totman 1989, 1992). By 628 A.D., 46 Buddhist monasteries were built. It has been estimated that 100,000 *koku* of processed lumber were used to build a single monastery (Totman 1989, p. 17). Tokoro, as cited by Totman (1989, p. 17), has estimated that the three centuries of monastery building that took place from 600 A.D. onwards consumed 10,000,000 *koku* of processed lumber.[8]

Continuous economic development also led to the export of timber to China, which by 1000 A.D. had started to import wood to meet its domestic needs. By 1000 A.D., the population of Japan had reached 6,500,000, and it was to double by 1600 A.D. Such surges meant that wood resources were required to meet the growing population. The capital city of Heian faded away by the twelfth century after its surrounding forests were cut (Totman 1992, pp. 19-20). To meet the socioeconomic expansion, Japan's old-growth forests were devoured by the seventeenth century, with continued monumental construction and rising growth in cities like Kyoto, Osaka, and Edo, whose populations ranged from 400,000 to 1 million persons. By 1720, the country's population had increased to 31 million.

Returning to the Mediterranean area at this time period, we find Venice as a center of accumulation, assuming the mastery of the seas with its superior ships. A giant ship factory was built, called the Arsenal, comprised of previously fragmented privately owned shipyards, which were organized into one state-operated facility. To satisfy the wood resources of this enterprise, the surrounding forests provided the wood and pitch for shipbuilding. The Venetian glass industry also required extensive wood, which led to deforestation of the nearby surrounding forests. To feed the growing population, pasture land had to be cleared. This pace of cut led to scarcity and, by 1530, shipbuilders had to pay twice the price for wood as their predecessors. As a consequence, the pace of shipbuilding had to be scaled back, thus affecting the accumulation processes. Furthermore, later in the century, Venetian ships were built in northern Europe, especially in Holland, thus giving the latter the opportunities for its ascent; it became a center of accumulation later during this period. The contest between Venice and the Ottoman Empire also engendered heavy deforestation to meet the shipbuilding needs of Venice. To Perlin (1989), this lack of access to wood further hampered Venice's position in the overall accumulation process and was one of the factors that led to its decline as major trading center (of accumulation). Venice's decline led to the shift of commercial power to northern Europe, which was then on the ascent.

By the fourteenth century, the forests of southern England had regenerated from the Roman smelting works of centuries before, and England was exporting wood to Holland and France. The growing rise of Holland as a center of accumulation meant that England was part of the peripheral hinterland supplying wood to fuel Holland's needs. Six hundred shiploads left England for France each year, and English wood, especially its oak, was in high demand (Perlin 1989, p. 163). England then had few industries requiring wood, as it exported natural resources and imported finished goods (necessities and luxuries) from the continent. However, later in the period, the English consumption of wood accelerated with the development of munitions production in southern England, around Sussex, under the reign of Henry VIII. With the production of iron for the manufacture of canon, wood consumption shot up, for a ton of bar iron required 48 cords of wood. Added to this consumption were the increased production processes occurring with respect to copper, salt, glass, and shipbuilding for the Royal Navy and merchant marine. For the latter, a large warship required 2,000 oak trees of at least a century old (Chew 1992, p. 25). Besides utilizing its own wood resources, England on its ascent to the core also exploited the forests of Ireland.

Ecological Degradation of the Period

Like the ancient civilizations of Mesopotamia and the Indus Valley, rapid deforestation in early China for agricultural production, irrigation, and the production of metallic implements and commodities generated conditions of ecological crisis. With population increases requiring more arable land, and the increasing utilization of wood to provide palaces, temples, tombs, and other personal pleasures of the elites, more pressure was added to already fragile ecological conditions. Such conditions began to be felt from the fifth century B.C. onwards (Bilsky 1980).

Deforestation throughout history has caused topsoil loss and siltation, and when forest cover is depleted, flooding usually follows (Roberts 1989; Ponting 1991).[9] Population increases, as we have previously noted, add pressure to increase agricultural production, which involves the clearing of land for cultivation. Also, wood as fuel for domestic use and manufacturing engenders deforestation. The limited statistics for early China in terms of population growth and the recorded number of floods in Chinese provinces reflect a high linear correlation (0.949) between population increases and the increasing number of floods (see Table 1). Regression analysis also shows that population increase is a good predictor of an increasing number of floods. The level of significance (r-square) is over 90 percent.

Placed within the wider context of the expansion and stagnation of growth over the long term, this suggests that environmental degradation also exhibits "long swings" correlating to population growth and economic expansion. The economic growth of China from 1 A.D. to the present reflects a logistic growth that correlates with population increases. The limited data suggest that there were four periods of logistic surges in population growth: from 400 B.C. to 200 A.D., 400 A.D. to 1200 A.D., 1550 A.D. to 1600 A.D., and 1700/1800 A.D. to the present. These population logistics are repeated for Europe over the same period.

Within these long swings of population growth, the number of floods over time also exhibits increases and decreases, with each phase lasting more than 200-300 years. Figure 1 and Table 2 outline the tempo of these long swings over the last two millennia, with the number of floods increas-

Table 1. Centennial Population of China
and Number of Floods

Year	Population (Millions)	Number of Floods
1AD	53.0	0
100	58.0	2
200	63.0	23
300	58.0	17
400	53.0	4
500	51.5	7
600	50.0	2
700	50.0	13
800	50.0	26
900	64.0	20
1000	66.0	42
1100	105.0	44
1200	115.0	72
1300	96.0	33
1400	81.0	81
1500	110.0	26
1600	160.0	52
1700	160.0	160
1800	330.0	172
1900	475.0	362

Sources: Chu (1926); Feuerwerker (1990).

Table 2. Number of Floods via A-B Phases
for China, 0-1700 A.D.

0-200	A-phase	Increase
200-400/500	B-phase	Decrease
400/500-750/800	A-phase	Increase
750/800-950	B-phase	Decrease
950-1200	A-phase	Increase
1200-1500	B-phase	Decrease
1500-1700	A-phase	Increase

ing and decreasing over the A-B phases of the system up to 1700 A.D.[10] The limited data suggest that the numbers of floods increase during A phases and slow during B phases. As the data are quite limited and are only for one country, further comparisons over time and different geographic space are required.

Conditions in China were repeated in Japan: erosion, extreme river silting, and flooding. Between 600-850 A.D., the mountains adjoining the Kinai basin were deforested, giving rise to wildfires, floods, and erosion. Deforestation was so extensive that by 675 A.D., the Emperor announced

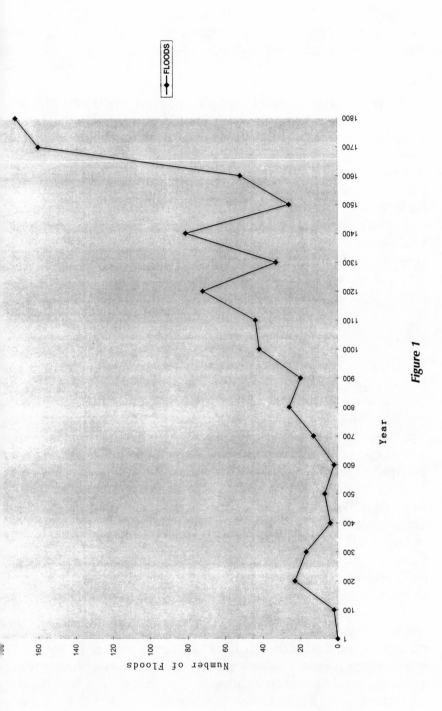

Figure 1

forest closure policies to protect the remaining stands of trees. This forest protection policy continued well into the ninth century A.D., although over-cutting continued to 1678, with its associated ecological consequences.

Besides ecological degradation, continuous assault on the landscape also engendered scarcity which, similar to previous periods such as in Hellenistic and Classical Greece, meant that production processes had to be relocated. For this period (500 A.D. to 1500 A.D.), English industrialists transported iron ore from England to Ireland for smelting because of the increasing scarcity of wood in England and the lower prices for it in Ireland. The price differential was about 50 to 100. Not only did core centers of accumulation have to relocate their production processes, but the lack of access to wood also engendered slippage in terms of their commercial and productive strengths. Venice's commercial and productive processes were affected by its lack of access to forest resources, and this affected its competition with the Ottoman Empire. Venetian transfer of shipbuilding to Holland facilitated the shipbuilding and commercial activities of Holland in northwestern Europe.

1500 A.D. to 1990 A.D.:   Asia, Europe, and North America

The rise of northern Europe in the world system following the fifteenth century led to increased utilization of wood for shipbuilding, construction, and manufacturing. Portugal, Spain, France, Holland, and England over different periods embarked on the exploitation of this resource. Portuguese penetration of the (Asian?) world system in 1498, landing first at Calicut and later at Goa and Malacca on the Malayan Archipelago, led to the cutting of trees to meet the needs of the Portuguese fleet (Gadgil 1988, p. 49). The Dutch colonization of Java in the seventeenth century engendered the cutting of teak wood to meet initial requirements of the Dutch East India Company, and later the imperial needs of Holland. Wood was sought in the initial stages for shipbuilding, building construction, furniture making, and charcoal fuel. Further consumption is revealed by the construction of shipyards, where by 1675 we find one that employed over 250 persons, and the presence of wind-powered sawmills. In concert with local rulers and middlemen traders, the Dutch East India Company harvested the teak trees which occupied about 6 percent of the total surface area of Java. Other trees, which had covered approximately 17 percent of Java, were deemed as junglewood and treated as worthless, suitable for destruction "of which nobody suffers" (Boomgaard 1988, p. 61). When the Dutch East India Company ceased to exist in 1800, control of the forests of Java was turned over to the Dutch Republic. In later years, especially in the nineteenth century, the degradation of the forests was exacerbated primarily to meet the production of exports such as sugar, coffee, indigo, and tobacco. This cash crop production required the increased felling of trees to generate arable land for cultivation, building construction, and fuel to process sugar, tobacco, and coffee. Annual production of teak grew from an average of 16,700 logs in 1733-1765 to 145,000 logs in 1837-1865 (Boomgaard 1988, pp. 66, 77).

Similar economic practices and forest policy also occurred in other parts of Southeast Asia where Great Britain was the colonial power. Malaya, Burma, and British Borneo witnessed the exploitation of their tropical forests by the British colonial administration to meet imperial economic ends. British companies searched for teak wood in Burma and, according to Rush (1991, p. 41), such quests "would play a part in Britain's annexation of Burma." Following the annexation of parts of Burma in 1826, British companies, supported by the British colonial administration, proceeded to exploit the wood resources (particularly, teak) of Burma and to clear the lands surrounding the Irrawaddy River for rice cultivation (Adas 1983). In Malaya, the forests were cut to grow rubber and open up areas to mine for tin. To facilitate the establishment of a rubber plantation economy, the British Government encouraged local rulers to sell their land and provided tax incentives to the new owners to plant rubber for export.[11] Similar to Java, the local Malay rulers, merchants, and traders were also intimately involved in the extraction of the forest and forest products. Borneo was administered by a chartered company, the British North Borneo Company, much like the Dutch East India Company which oversaw the management and harvesting of trees. The forest and forest products were, in many cases, the main source of revenue for the British North Borneo Company (Lian 1988). In Thailand, which was not colonized by the British, British timber companies also penetrated teak forests in the nineteenth century.[12] In the early twentieth century, Thai forests were granted as concessions to foreign companies from Great Britain, France, and Denmark.

This manner of economic activity and penetration occurred also for the Philippines, except that it was colonized first by Spain in the sixteenth century and, later, by the United States (in the late nineteenth century). Spanish colonization saw the island of Cebu stripped of its hardwood to build Spanish galleons (Tadem 1990, p. 15). Under American rule, the Forest

Law of 1904 formed the blueprint for the modernization of the logging industry of the Philippines and cemented the close relationship between the Bureau of Forestry and large foreign (American, British, and Spanish) and domestic timber companies. A number of other laws, such as the Public Lands Act of 1902, the Mining Law of 1905 and Executive Order No. 27 of 1929, also facilitated concentration of private ownership of land and opened up the forests for private commercial exploitation. Huge profits were made from the harvest of the tropical forests of the Philippines (Tucker 1988, p. 223). As with other parts of Southeast Asia, the local elites were active participants in this accumulation process. Such a heightened pace of activity led the transformation of the Philippines from a wood importer to a wood exporter as a direct result of American colonial policy. The pace continued, and the exploitation of the country's natural resources by foreign concerns (especially from America) transformed the economy to one dependent on natural resource exports to the United States (Bellow et al. 1982). Annual rate of deforestation was at least about 140,600 hectares per year during the 1920s and 1930s. This rate increased slightly by another 30,000 hectares per annum in the post-independence era (Bautista 1990, p. 69).

Back in Europe, English merchants and the state sought to replenish England's wood supply, which was running low by the mid-seventeenth century (Chew 1992). This condition had been reached as a result of the previous century of growth—the "long sixteenth century" of expansion. Wood was used extensively in the construction of buildings and in manufacturing. It was, as some would say, an "age of wood" (Lower 1973). Tanners, soap boilers, gunpowder manufacturers, and glass makers all required wood products, such as oak bark and potash, for their operations. The mining industry needed sturdy timber for its mine shafts. Brick manufacturers, sugar refiners, and salt producers used wood. For example, 50 cubic feet of hardwood was required to produce 2,000 bricks, and more than a load of hardwood was needed to produce two hundredweights of salt (Chew 1992, p. 23). Such intensive utilization of wood for manufacturing operations was exacerbated further by the enclosure movements and the intensive growth in farming prior to the mid-seventeenth century, which led to an extensive cutting of trees. We find England seeking out wood resources in two continents: northeastern Europe and North America. With the increasing need to find an overseas supply, the forests of these areas became targets for assault. Thus, the fir forests of the Baltic shores and the oak belt found to the south and west, along with the forests of New

England, became important regional sources of British timber supplies. At this stage, the Baltic area was preferred because of its proximity and, thus, lower transportation costs, as a source of wood.[13] However, with the Continental blockade proclaimed in 1807 by Napoleon, British timber interests shifted their production operations to North America.[14] The pace of operations continued throughout the nineteenth century, and along with indigenous American timber and lumber operations, the forests on the eastern seaboard of North America continued to be assaulted until the early parts of the twentieth century. Deforestation occurred, though these areas had lower population densities and, thus, were less impacted in terms of floods and other consequences. After the turn of the twentieth century, the exploitation of the forests of North America progressively moved to the American Midwest and West Coast. Wood exports of primarily Douglas fir and coastal redwoods were shipped from the Pacific Northwest.

In the late twentieth century, the exploitation of the forests continues to be a global process. The assault occurs across Europe, Africa, North America, Asia, and Latin America. In Latin America and Asia, the tropical rainforests have been systematically cut to meet accumulation and consumption needs, which are both local and global in nature. Operating either independently or in collaboration with local timber and lumber concerns, multinational companies based in Japan, Canada, and the United States have been active in harvesting the forests (Petesch 1990). In India, the forests of the Himalayas are being assaulted (Guha 1989). In Southeast Asia, Japanese *sogo shoshas*[15] have managed to penetrate the tropical rainforests (Chew 1993). Debt servicing, multinational activities, ranching, and international organizations have been responsible for the destruction of the Brazilian rainforests and other forests in Central and Latin America (Barbosa 1993; Hecht and Cockburn 1990). With mounting environmental group pressures in North America, the search for wood has even moved to Siberia, and Canada's boreal forests are now seen as potential resource bases (Goto 1993, p. 6; Chew 1993). Japanese multinationals, such as Mitsubishi and Sumitomo in conjunction with Svetlaya, are in a joint venture with the Hyndai Corporation of Korea to exploit the Siberian taiga (Goto 1993, p. 6).

In the late twentieth century, the United States and Japan have replaced England and France as the two main centers of accumulation and, consequently, of wood extraction. At this stage, the dynamics of wood exploitation occur in parts of the world system that are the most amenable to the accumulation of capital. Consequently, we find extensive wood operations by American and Japanese multinationals in North America, Latin America, Asia, Siberia, and West Africa. For Japan, whose wood resources were severely exploited in the past, global efforts are being made now by its *sogo shoshas* to maintain a constant supply from foreign sources (Nectoux and Kuroda 1990, p. 27). Japanese dependency on imported wood has risen from 5.5 percent in 1955 to 55 percent in 1970 and 66.5 percent in 1986 (Nectoux and Kuroda 1990, p. 27). With a more bountiful indigenous wood supply, the United States has been more fortunate than Japan. However, this has not stopped multinational operations of U.S.-based corporations from exploiting the temperate and tropical forests in the periphery of the world system.

Core-periphery relations have also engendered differential accumulation rates and fostered the ascendancy of core states in the world system, as evidenced by the history of accumulation via timber and lumber operations. In the nineteenth century, England along with France seemed to have benefitted. In the late twentieth century, besides the United States, we find Japan exhibiting a global accumulation strategy in the timber and lumber sector of the world economy. The activities of Japan's *sogo shoshas*, including wood import levels, reflect the intensiveness and breadth of the capital accumulation process. For nearly two decades, Japan has been the world's major tropical timber hardwood importer. The total volume of tropical wood imports into Japan amounted to 29 percent of world trade in tropical hardwoods in 1986 (Nectoux and Kuroda 1990, p. 5). The high levels of consumption reflect an age of exuberance much like that which the United States experienced in the postwar era (Devall 1993).[16] Regarding paper products, over the last 30 years, Japan has emerged as the world's largest importer of forest products, the second largest producer of paper and paperboard, the third largest producer of pulp, and the second largest consumer of paper in the world after the United States (Penna 1992, p. 1). Total consumption of paper and paperboard increased over 548 percent during the last 30 years (Penna 1992, p. 3).

Because of its proximity to Southeast Asia, Japan has treated the region as its own wood yard, just as England did with British North America in the eighteenth to nineteenth centuries. Over the last decade, Japan has been importing unfinished logs from Southeast Asia on an average of 11 million cubic meters, or about 38 percent of its total volume of log imports (Kato 1992, p. 95). Malaysia, Indonesia, and Papua New Guinea were the main contributors (Mori No Koe 1993, p. 8). By the end of the 1980s, this rate slowed, not because of a drop in consumption but because of a diminishing resource and the bans enforced by the states in Malaysia, Indonesia, and the Philippines on the export of raw logs.[17] Other contributing countries to the wood industry of Japan were Canada, the United States, the former Soviet Union, and New Zealand.

What does this mean for the Japanese corporations involved in this sector of the world economy? Investments in the pulp and paper industry have grown steadily, from 1.8 billion yen in 1980 to 4.7 billion yen in 1989 (Graham 1993, p. 39). This includes direct investments in other parts of the world economy, such as the United States, Canada, Chile, Indonesia, Thailand, Malaysia, Portugal, New Zealand, Australia, and Papua New Guinea (Penna 1992, pp. 29-30). According to Penna (1992, p. 18) the concentration of production share in the industry has increased dramatically over the last 30 years. In 1960, the top 10 Japanese pulp and paper corporations produced 61 percent of the pulp, 64 percent of the paper, and 42 percent of the paperboard in Japan. By 1990, the top 10 corporations accounted for the manufacture of 83 percent of the pulp, 74 percent of the paper, and 57 percent of the paperboard. Such a high volume of consumption naturally leads to an immense power on the part of Japan's pulp and paper industry over its peripheral producers in the world, such as in Southeast Asia. Tadem (1990, p. 23) has indicated that Japan has been labeled an economic imperialist by some Southeast Asia wood producers, for it levies higher import taxes on finished wood products than on logs. For example, a 20 percent import tax is levied on plywood imports, as compared to zero percent on unprocessed logs. It has been suggested that this measure protects Japan's labor-intensive wood manufacturing industry (Tadem 1990, p. 24). Attempts by Southeast Asia producers (Council of Southeast Asian Lumber Producers Association) to control supply have been met with Japan's slashing the prices of logs and lumber "to break up any wood and forest product cartel in the region" (Tadem 1990, p. 25). Such actions are hardly unique in the history of the world system. One need only look at past British and American policies toward their colonies or those countries under their scope of economic penetration and domination. No doubt, the search for wood products to fuel the processes of production and reproduction will continue on a global level as long as the world system is organized around ceaseless accumulation as its motive force.

Ecological Degradation of the Period

If we move forward to the late twentieth century, with deforestation at a quickened pace as a result of improvements in timber harvesting technology, disastrous effects can be seen throughout the whole of North America, Latin America, and Asia, with massive "clear-cuts" (Devall 1994; Chew 1993; Mendez 1990). The consequences of this deforestation correspond to other zones and other periods of the world system: river siltation, soil erosion, flooding, and certain animal, insect, and plant species extinctions. For example, in the American Pacific Northwest, several ecosystems are being threatened, and a biodiversity crisis is emerging (Grumbine 1992). In Asia, the most dramatic deforestation occurred between 1950 to 1976, according to the United Nations Economic and Social Commission for Asia and the Pacific (1980), when 4 million hectares were cleared per annum. Soil loss, erosion, and intermittent flooding of lowland areas during the monsoon season have been consequently experienced in Malaysia, Indonesia, Thailand, the Philippines, India, and the Himalayas. Soil loss is followed by sedimentation problems in terms of river flows and harbor depths. This has often led to flooding, as in West Malaysia, and harbor dredging, as in Sarawak in East Malaysia. Indonesia, the Philippines and Thailand have also experienced these conditions. Concomitant are the effects on fish stocks, either in the mangrove areas of Malaysia, Indonesia, and the Philippines or in the Gulf of Thailand. Sedimentation has also affected the pH levels of the rivers and, in turn, this has affected aquaculture and fishery stocks in the coastal zones.

Besides erosion, the other urgent and important threat to the environment as a consequence of deforestation is the loss of biodiversity. The tropical rainforests of Southeast Asia have one of the greatest diversity of plant and animal species on the planet. The loss to date has been drastic. In West Malaysia, which has 7,900 species of flowering plants, 207 species of mammals, 495 species of birds, and 250 species of freshwater fish, logging has placed almost 61 species of mammals and 16 species of birds on the verge of extinction, with a further 130 species of mammals and 148 species of birds on the vulnerability threshold. In terms of number of primates lost as a result of deforestation, Hurst (1990, p. 121) indicates that some species, such as the Siamang, the White-Handed Gibbon, and the Dusky leaf-monkey, suffered population losses of over 50 percent between 1958 and 1975. No doubt these losses have increased by now. The threat to wildlife diversity also affects Indonesia, Thailand, and the Philippines. The Philippine Monkey-Eating Eagle has been reduced to only 600 mating pairs, and the elephant population in Thailand and the orangutan and Java rhino in Indonesia are threatened.

CONCLUSION

This examination of the dynamics of world accumulation vis-à-vis the production and consumption of timber and wood products underscores the intertwining of socioeconomic processes and the rhythms of nature. It also outlines the instability and vulnerability circumscribing these relations. Our brief review of world history has shown this recurring dynamic whereby nature has been shaped macroparasitically to meet the dynamics of the reproduction of social organizations that have existed and perished in some cases. The historical journey has provided some suggestive accounts of the continuity and the self-defeating nature of the accumulation process over time-space dimensions with the ecological degradation that ensued. Ecological degradation has been a continuous process over world history, though not necessarily in the same geographic regions over time. With the rise and fall of centers of accumulation, ecological degradation does shift from one region to another depending on where the centers of accumulation are located in the world system. It is also clear that some regions do not fully recover from the devastation if the assault on the environment is ceaseless. For example, witness the state of the forests today in what was formerly Roman North Africa a millennium years ago, devastated not only by the Romans but, as well, by those who followed, such as the Spanish, the Italians, the English, the French, and others.

Despite the continuity of the historical process of accumulation, nature in the long run still defines the parameters and conditions of production and accumulation in world history. Lack of access to natural resources such as wood, or its scarcity, has reduced the competitive edge for some centers of accumulation in the history of the world system. It has also been suggested by some scholars that kingdoms and civilizations have collapsed due to the degradation of their environments. To place environmental degradation as the sole cause of collapse might be too extreme. Rather, we should consider it as part of a set of factors contributing to the demise. In light of this, the natural dimension and its limits for global transformation should be realized, especially for analyses that rely on a materialistic view.

Realizing the continuity of the ecological crisis, with its limits on capital accumulation processes, suggests that as the accumulation process intensifies, environmental degradation increases exponentially. The limited long-term data on environmental degradation in China (using flooding as an example) do suggest that there might be "long swings" in environmental degradation correlating with the long-trend logistical increase in population and economic growth over world history, and also with expansion and contraction phases of long economic cycles. Further investigations over space-time dimensions might help to clarify this.

Furthermore, with environmental crisis-like conditions pervading world history, these factors also spark societal ecological consciousness and conservation practices (Chew 1995). Viewed through this long-term perspective, the current concern and debate over environmental crises reflect similar practices of prior historical periods of the world system. The question is whether our current environmental crisis is one which is qualitatively different from prior periods of world history. Such an answer cannot be easily provided due to the limited materials and information available over long periods of world history for assessing the impact of environmental crisis as a consequence of the world accumulation process. It appears that with the "cumulation of accumulation" over world history, the continuity of the assault of the environment has reached exponential levels. With the current global reach of capital, at this point in world system history the possibility of a *global* environmental devastation may be more likely than before.

ACKNOWLEDGMENTS

Thanks to Andre Gunder Frank, Al Bergesen, Bill Devall, Pat Lauderdale, and Jan Tye-Chew for their helpful comments on an earlier draft. An earlier version of this paper was presented to the Meeting on World System History: The Social Science of Long Term Change, University of Lund, Sweden, March 25-28, 1995.

NOTES

1. Abu-Lughod (1989) has viewed these tribute missions as "public" trading.
2. Wheatley (1961, p. 17) has indicated that this state was peopled by families belonging to the Mongol and Tartar tribes of Central Asia.
3. Wang (1958, pp. 124-125) has indicated that Persian merchants had contacts with China as early as 518 A.D., but this was by land through the Silk Route.
4. Abu-Lughod sees it as an interplay of three interlocking circuits (stretching from the Arabian peninsula to southern China), each "within the shared 'control' of a set of political and economic actors who were largely, although certainly not exclusively, in charge of exchanges with adjacent zones" (1989, pp. 251-253).
5. This point was raised by Gunder Frank in our personal conversation on the role of Southeast Asia in world system history.
6. The state was so supportive that it awarded yearly renewal of trading licenses to Chinese merchants who returned with cargoes dutiable at more than 5,000 strings of cash.

Hotel accommodations were provided to foreign traders, and welfare benefits to foreign shipwrecked seamen (Wheatley 1959, p. 25).

7. There have been a number of conjectures on why China withdrew from this area (see Abu-Lughod 1989).

8. In terms of volume of wood utilization, 100,000 *koku* of processed lumber is sufficient to build 3,000 ordinary 1950s Japanese-style units of housing (Totman 1989).

9. It should be noted that floods are not caused only by deforestation; other factors engendering flood conditions, such as temperature changes and climatological shifts, can also account for floods.

10. The slicing of A-B phases over world history is derived from Frank (1993).

11. By the 1950s, tin and rubber were the major exports of Malaya and covered approximately 15,000 sq. km of the country.

12. The penetration of Thailand was made during the reign of King Mongkut, who signed the Bowring Treaty with British companies. The teak was paid for with opium (Rush 1991, p. 16).

13. For a detailed discussion of the political economy of timber and lumber production in this area, see Albion (1926) and Chew (1992).

14. For a detailed account of the shift in production operations to British North America, see Chew (1992).

15. The *sogo shosha* is a trading house which provides financing for timber operations in the peripheral areas, credit for processing, and sales. There are approximately 8,500 of these trading houses in Japan; however, the 15 largest ones handle the majority of timber imports into Japan (Nectoux and Kuroda 1990; Penna 1992).

16. It should be noted that due to mounting environmental group pressure in North America, a more consistent supply of wood resources is now being sought in Siberia (Goto 1993, p. 6). Since the 1960s, Japan has imported wood from the former Soviet Union, though Soviet wood could not compete against wood from North America in terms of quality and the Soviets had an antiquated production process marred by outdated machinery.

17. Indonesia started to restrict the export of logs in 1979, while Malaysia took almost more than a decade later to ban the temporary export of logs in January 1993 and to lift the ban with quota level exports in May 1993 (Mori No Koe 1993, p. 6).

REFERENCES

Abu-Lughod, J. 1989. *Before European Hegemony*. New York: Oxford University Press.

Adas, M. 1983. "Colonization, Commercial Agriculture and Destruction of the Deltaic Rainforests of British Burma in the Late 19th Century." Pp. 95-110 in *Global Deforestation and the 19th Century World Economy*, edited by R. Tucker and J. Richards. Durham, NC: Duke University Press.

Albion, R.G. 1926. *Forests and Seapower: The Timber Problem of the Royal Navy*. Cambridge. MA: Harvard University Press.

Amin, S. 1974. *Accumulation on the World Scale*. New York: Monthly Review Press.

Attenborough, D. 1987. *The First Eden: The Mediterranean World and Man*. Boston: Little Brown.

Barbosa, L. 1993. "The World-System and the Destruction of the Brazilian Amazon Rainforest." *Review* XVI(2): 215-240.

Bautista, G. 1990. "The Forestry Crisis in the Philippines: Nature, Causes and Issues." *The Developing Economies* XXVIII(1, March): 67-94.

Bellow, W. et al. 1982. *Development Debacle: The World Bank in the Philippines.* Berkeley: University of California Press.

Bilsky, L. 1980. "Ecological Crisis and Response in Ancient China." Pp. 60-70 in *Historical Ecology,* edited by L. Bilsky. New York: Kennikat Press.

Boomgaard, P. 1988. "Forests and Forestry in Colonial Java." Pp. 59-87 in *Changing Tropical Forests,* edited by J. Dargavel. Canberra: Center for Resource and Environment Studies.

Butzer, K.W. 1971. *Environment and Archaeology from an Ecological Perspective.* Chicago: Aldine-Atherton.

Chandler, T. 1974. *3,000 Years of Urban Growth.* New York: Academic.

Chase-Dunn, C., and T. Hall, eds. 1991. *Core-Periphery Relations in Pre-Capitalist World.* Boulder, CO: Westview Press.

Chase-Dunn, C., and T. Hall. 1996. *Rise and Demise: Comparing World-Systems.* Boulder, CO: Westview.

Chaudhuri, K.N. 1985. *Trade and Civilization in the Indian Ocean: An Economic History from the Rise of Islam to 1750.* Cambridge: Cambridge University Press.

Chew, S.C. 1992. *Logs for Capital The Timber Industry and Capitalist Enterprise in the 19th Century.* Westport, CT: Greenwood Press.

_____. 1993. "Wood, Environmental Imperatives and Development Strategies: Challenges for Southeast Asia." Pp. 206-226 in *Asia—Who Pays for Growth? Women, Environment and Popular Movements,* edited by J. Lele and W. Tettey. London: Dartmouth.

_____. 1995. "Environmental Transformations: Accumulation, Ecological Crisis, and Social Movements." Pp. 201-216 in *A New World Order? Global Transformation in the Late Twentieth Century,* edited by D. Smith and J. Borocz. Westport, CT: Greenwood Press.

Chu, C-C. 1926. "Climatic Pulsations During Historic Time in China." *Geographical Review* 16(2): 274-282.

Coedes, G. 1983. *The Making of S.E. Asia.* Berkeley: University of California Press.

Colless, B.E. 1969. "Persian Merchants and Missionaries in Medieval Malaya." *Journal of the Malayan Branch of the Royal Asiatic Society* 42(2): 10-47.

Devall, B. 1993. *Living Richly in an Age of Limits.* Salt Lake City, UT: Peregrine Smith.

_____. 1994. *Clear Cut the Tragedy of Industrial Forestry.* San Francisco: Sierra Club Books.

Dunn, F.L. 1975. *Rain Forest Collectors and Traders: A Study of Resource Utilization in Modern and Ancient Malaya #5.* Kuala Lumpur: Malayan Branch of the Royal Asiatic Society.

Edens, C. 1990. "Indus-Arabian Interaction During the Bronze Age." Pp. 335-361 in *Harappan Civilization and Rajdi,* edited by G. Possehl. New Delhi: Oxford University Press.

Ekholm, K., and J. Friedman. 1982. "'Capital' Imperialism and Exploitation in Ancient World-Systems." *Review* IV(1): 87-109.

Feuerwerker, A. 1990. "Chinese Economic History in Comparative Perspective." Pp. 224-241 in *Heritage of China: Contemporary Perspective on Chinese Civilization,* edited by P. Ropp. Berkeley: University of California Press.

Frank, A.G. 1991. "Transitional Ideological Modes: Feudalism, Capitalism, Socialism" *Critique of Anthropology* 11(2): 171-188.

———. 1992. "The Five Thousand Year World System: An Interdisciplinary Introduction." *Humboldt Journal of Social Relations* 18(2): 1-80.

———. 1993. "Bronze Age World System Cycles." *Current Anthropology* 34(4): 383-405.

———. 1994. "The World System in Asia Before European Hegemony." *The Historian* 56(4): 259-276.

Friedman, E. 1982. *Ascent and Decline in the World-System*. Beverley Hills, CA: Sage.

Friedman, J. 1992. "General Historical and Culturally Specific Properties of Global Systems." *Review* XV(3): 335-372.

Gadgil, M. 1988. "On the History of the Kannada Forests." In *Changing Tropical Forests*, edited by J. Dargavel. Canberra: Centre for Resource and Environmental Studies.

Gadgil, M., and Guha, R. 1992. *This Fissured Land: An Ecological History of India*. Berkeley: University of California Press.

Gills, B., and A.G. Frank. 1992. "World System Cycles, Crises, and Hegemonial Shifts 1700 BC to 1700 AD." *Review* XV(4): 621-687.

Goto, D. 1993. *Logging in Siberia: Japan's Involvement. Japan Environmental Exchange Newsletter* (Tokyo): July.

Graham, A. 1993. *Wood Flows Around the Pacific Rim (A Corporate Picture)*. Paper presented at the First International Temperate Forest Conference, Tasmania, Australia.

Grumbine, E. 1992. *Ghost Bears*. Covelo, CA: Island Press.

Guha, R. 1989. *The Unquiet Woods*. Berkeley: University of California Press.

Hecht, S., and A. Cockburn. 1990. *The Fate of the Forest*. New York: Harper.

Hoffman, M. 1980. "Pre-Historic Ecological Crises." Pp. 33-42 in *Historical Ecology*, edited by L. Bilsky. New York: Kennikat Press.

Hurst, P. 1990. *Rainforest Politics*. London: Zed Press.

Jacobsen, T., and R.M. Adams. 1958. "Salt and Silt in Ancient Mesopotamian Agriculture." *Science* 128: 1251-1258.

Kato, T. 1992. "Structural Changes in Japanese Forest Products Import During the 1980s." Pp. 87-101 in *The Current State of Japanese Forestry*. Tokyo: Japanese Forest Economic Society.

Kohl, P. 1987. "The Ancient Economy, Transferable Technologies, and the Bronze Age World System: A View from the Northeastern Frontier of the Ancient Near East." Pp. 13-24 in *Centre and Periphery in the Ancient World*, edited by M. Rowlands, M. Larsen and K. Kristiansen. Cambridge: Cambridge University Press.

Ladurie, E.L.R. 1971. *Times of Feast, Times of Famine: A History of Climate Since the Year 1000*. New York: Doubleday.

Lian, F. 1988. "The Economies and Ecology of the Production of the Tropical Rainforest Resources by Tribal Groups of Sarawak, Borneo." Pp. 34-44 in *Changing Tropical Forests*, edited by J. Dargavel. Canberra: Centre for Resource and Environmental Studies.

Lim, H.F. 1992. "Aboriginal Communities and the International Trade in non-Timber Forest Products: The case of Peninsular Malaysia." Pp. 77-88 in *Changing Pacific Forests*, edited by J. Dargavel and R. Tucker. Durham, NC: Forest History Society.

Lo, J.P. 1958. "The Decline of the Ming Navy." *Oriens Extremus* V: 149-168.

Lower, A. 1973. *Great Britain's Woodyard*. Kingston: Queen's University Press.

Marsh, G. 1965. *Man and Nature: Or Physical Geography*. New York: Charles Scribner.

Marshall, J. 1931. *Mohenjo-daro and the Indus Civilization,* 3 vols. Cambridge: Cambridge University Press.

McNeill, W. 1992. *The Global Condition.* Princeton, NJ: Princeton University Press.

Meiggs, R. 1982. *Trees and Timber of the Ancient Mediterranean World.* Oxford: Clarendon Press.

Mendez, C. 1990. *Fight for the Forest.* London: Latin American Bureau.

Modelski, G., and W. Thompson. 1996. *Leading Sectors and World Powers: The Coevolution of Global Economics and Politics.* Columbia: University of South Carolina Press.

Mori No Koe. 1993. *Japan and the World's Forests.* Tokyo.

Nectoux, F., and Y. Kuroda. 1990. *Timber from the South Seas.* Geneva: World Wildlife Fund.

Penna, I. 1992. *Japan's Paper Industry.* Tokyo: Chikyu no Tomo.

Perlin, J. 1989. *A Forest Journey: The Role of Wood in the Development of Civilization.* Cambridge, MA: Harvard University Press.

Petesch, P. 1990. *Tropical Forests: Conservation with Development.* Washington, DC: Overseas Development Council.

Ponting, C. 1991. *A Green History of the World.* New York: St. Martin's Press.

Randsborg, K. 1991. *The First Millennium A.D. in Europe and the Mediterranean: An Archaeological Essay.* Cambridge: Cambridge University Press.

Ratnagar, S. 1981. *Encounters: The Westerly Trade of the Harappan Civilization.* New Delhi: Oxford University Press.

Roberts, N. 1989. *The Holocene: An Environmental History.* Oxford: Basil Blackwell.

Rush, J. 1991. *The Last Tree: Reclaiming the Environment in Tropical Asia.* New York: Asia Society.

Shaw, B. 1981. "Climate, Environment, and History: The Case of Roman North Africa." Pp. 379-403 in *Climate and History,* edited by T. Wrigley. Cambridge: Cambridge University Press.

Tadem, E. 1990. "Conflict over Land Based natural Resources in the ASEAN Countries." Pp. 13-50 in *Conflict over Natural Resources in S.E. Asia and the Pacific,* edited by T.G. Lim and M. Valencia. Singapore: Oxford University Press.

Thirgood, J.V. 1989. *Man and the Mediterranean Forest: A History of Resource Depletion.* London: Academic Press.

Thomas, K. 1983. *Man and the Natural World: A History of Modern Sensibility.* New York: Pantheon Books.

Tibbetts, G.R. 1956. "Pre-Islamic Arabia and S.E. Asia." *Journal of the Malayan Branch of the Royal Asiatic Society* 29(3): 182-208.

_____. 1957. "Early Muslim Traders in S. E. Asia." *Journal of the Malayan Branch of the Royal Asiatic Society* 30(Part 1): 1-45.

Totman, C. 1989. *The Green Archipelago.* Berkeley: University of California Press.

_____. 1992. "Forest Products Trade in Pre-Industrial Japan." Pp. 19-24 in *Changing Pacific Forests,* edited by J. Dargavel. Durham, NC: Forest History Society.

Tucker, R. 1988. "The Commercial Timber Economy under Two Colonial Regimes in Asia." Pp. 23-35 in *Changing Tropical Forests,* edited by J. Dargavel. Canberra: Centre for Resources and Environmental Studies.

United Nations Economic and Social Commission for Asia and the Pacific. 1980. *State of the Environment for Asia and the Pacific.* Bangkok: U.N.

Wallerstein, I. 1974. *The Modern World-System,* Vol. 1. San Diego, CA: Academic Press.
_____. 1991. "World System versus World-Systems: A Critique." *Critique of Anthropology* 2(2): 189-194.
_____. 1992. "The West, Capitalism, and the Modern World-System." *Review* 15(4): 561-620.
Wang, G. 1958. *The Ninhai Trade: A Study of the Early History of Chinese Trade in S.E.Asia. Journal of the Malayan Branch of the Royal Asiatic Society* 31(2): 1-135.
Wheatley, P. 1959. *Geographical Notes on Some Commodities Involved in the Sung Maritime Trade. Journal of the Malayan Branch of the Royal Asiatic Society* 32(2): 1-140.
_____. 1961. *The Golden Khersonese.* Kuala Lumpur: University of Malaya Press.
Wheeler, R. 1968. *The Indus Civilization.* Cambridge: Cambridge University Press.
Wilkinson, D. 1991. "Core, Peripheries and Civilizations." In *Core-Periphery Relations in the Ancient World,* edited by C. Chase-Dunn and T. Hall. Boulder, CO: Westview Press.
Wolters, O. 1967. *Early Indonesian Commerce.* Ithaca, NY: Cornell University Press.
_____. 1982. *History, Culture and Region in Southeast Asian Perspectives.* Singapore: Institute of Southeast Asian Studies.
Yamamoto, T. 1981. "Chinese Activities in the Indian Ocean Before the Coming of the Portuguese." *Diogenes* 111: 19-34.

DYNAMICS AND RHETORICS OF SOCIO-ENVIRONMENTAL CHANGE: CRITICAL PERSPECTIVES ON THE LIMITS OF NEO-MALTHUSIAN ENVIRONMENTALISM

Peter J. Taylor and Raúl García-Barrios

ABSTRACT

We argue that environmental degradation is not related in any direct way to population numbers or growth rate. We focus on the poor in rural societies and show that one needs to analyze the dynamics linking changes in the labor supply, the social organization of production, technology, and the environment. Implicated in the maintenance, breakdown, or reorganization of local institutions of production are the differentiation in any society or community, its social psychology of norms and reciprocal expectations, and larger economic structures. In contrast, what we call neo-Malthusian environmentalism points to aggregate regional, national, or global statistics and to calculations of ultimate biophysical limits. We argue that these statistics and calculations give very little insight into the social-economic-environmental dynamics of socio-environmental change.

Advances in Human Ecology, Volume 6, pages 257-292.

257

Nevertheless, both the science and the politics of neo-Malthusian environmentalism have persistent appeal. We interpret this in terms of rhetoric privileging moral and technocratic conceptions of social action. The consequences of this rhetoric are unintended and undesirable effects which, contrary to the intentions of most environmentalists, contribute to coercion and violence in the name of the environment.

The conceptual and empirical weaknesses we expose in neo-Malthusian environmentalism and the implications we draw from its rhetoric are intended to challenge all environmentalists and human ecologists who want to promote sustainability and equity.

INTRODUCTION

Sustainable development, steady-state economics, and zero population growth are intended to be serious proposals. The economic and environmental problems motivating these goals are severe, and the social and economic changes their implementation seems to require are sweeping. Yet, the proponents of such steady-state or sustainable society goals often picture the dynamics of unsustainability, economic growth, and population increase very simplistically. Aggregated categories and abstract analyses of statistical trends predominate over investigations of concrete social, economic, and environmental dynamics involving unequal and *differentiating* agents. We believe that policies and other social or technical practices are more likely to succeed *without unintended and undesirable effects* if they are based on a sufficient description of the causes underlying such dynamics. Any sustainable social order will, after all, have to be constructed through interventions within these dynamics (Max-Neef 1986). Serious conceptual and empirical work to understand those dynamics is needed.[1]

At the same time, we recognize that simplistic or poorly framed analyses do not just happen spontaneously. The sociology of scientific knowledge indicates that certain courses of action are facilitated over others in the very formulation of science, that is, not just in its "downstream" applications. If our analysis is to shift the direction of analysis, policymaking, and other action, in addition to exposing conceptual and empirical weaknesses, we need to advance some interpretation that exposes the bases of support for the science and politics behind any steady-state proposals. Ideally, this would then help readers contribute to building conditions favorable to alternative science and politics. That last project, however, requires much more work than one written intervention can accomplish; here, we concentrate on interpreting simplistic analyses in terms of rhetoric privileging "undifferentiated" conceptions of social action.

In order to make concrete the directions we think such analysis of dynamics and interpretation of rhetoric should take, this paper focuses on one form of steady-state environmental discourse, what we call neo-Malthusian environmentalism and, in particular, on its account of the interrelations between the poor in rural societies and their environments.[2]

Positioning This Critique

Upon entering this terrain, one is quickly faced with contested definitions of who and what constitute neo-Malthusianism, popular slogans concerning the global and local, statements of self-evident truths about the finiteness of the earth, and the well-guarded disciplinary turf of demography. Let us, therefore, clear some ground for ourselves by defining some terms, suggesting several distinctions, and thereby providing a basic map of the surrounding area.

For us, neo-Malthusianism means more than a focus on over-Population and Population control; we shall refer to that focus as Population discourse or the Population problem. (From this point on, Population will be capitalized when it is used in a numerical and demographic, rather than sociological, sense.) We use the term neo-Malthusianism when ultimate biophysical limits, often global, are invoked to strengthen claims that Population growth presents a serious problem, one that should be kept at the center of our attention. When degradation of environments and exhaustion of resources are directly related to such Population growth, we call this neo-Malthusian environmentalism (e.g., Ehrlich and Holdren 1971; Ehrlich and Ehrlich 1990; Bongaarts 1992; Meffe et al. 1993; Hall et al. 1994; Pimentel et al. 1994).

For the purposes of this paper, demography—the scientific discipline spawned by and now dominating Population discourse—can be divided into three orientations (Preston 1989): (1) macroeconomic (concerned with the effects of Population growth on a nation's production and economic growth); (2) microeconomic (concerned with allocating the true social and economic costs of having children to those who bear them); and (3) reproductive/contraceptive practices (concerned with explaining different practices and, with the more recent emphasis on reproductive health and choice, of enabling mothers, and sometimes also fathers, to have the number of children they desire and raise them heathily). Each of these orientations may be developed with a neo-Malthusian tone. It is macroeconomic considerations, however, that have been most commonly associated with neo-Malthusian environmentalism, and so our discussion will speak most directly to that orientation.[3]

Global change is a very popular term these days, but, with a view both to identifying causes and to designing policy responses, we consider global formulations to be weak and unhelpful.[4] Global statistics and trends, or, more generally, aggregate regional or national figures, are abstractions that give very little insight into concrete social-economic-environmental processes (Palloni 1994).[5] Whatever the scale of observation, differentiation among social groups is at the center, not just an addition to, all such processes.[6] Let us tease out this assessment of undifferentiated thinking, which includes global formulations, with a simple, but powerful, scenario.

Consider two hypothetical countries having the same amount and quality of arable land, the same population size, the same level of technical capacity, and the same population growth rate, say, 3 percent per year. Country A, however, has a relatively equal land distribution, while country B has a typical 1970s Central American land distribution: 2 percent of the people own 60 percent of the land; and 70 percent own just 2 percent. Both countries double their Populations very rapidly but five generations (120 years) before anyone is malnourished in country A, all of the poorest 70 percent in country B already are. This is not just an issue of relative timing of the crisis in the two countries. The likely level at which B's poor would first experience what others call Population pressure would be food shortages linked to inequity in land distribution (see Durham 1979; Vandermeer 1977). Inevitably, given that no real country is like country A, the crises to which actual people have to respond come well before and *in different forms from* the crisis predicted on the basis of the aggregate Population growth rates and calculations of ultimate biophysical limits. Anyone focusing on Population control policies could justifiably be viewed by the poor in a country such as B as taking sides with those who benefit from the inequitable access to productive resources. The point of the scenario is not just that in any district, country, or ecosphere, there are strata of richer and poorer people, but that groups with different wealth and power exist, change, and become involved in crises because of their dynamic interrelations.

From this scenario, we identify three analytic/policy orientations, differing in the units of analysis (the kind of person or other agent who is involved in the phenomenon) and the implied limit (that makes the phenomenon a problem): (1) uniform, undifferentiated units (which can be simply aggregated) with biophysical limits; (2) unequal, stratified units, where the economic squeeze on the poor leads them to face biophysical limits, while the rich, buffered for some time from such limits, and can take anticipatory action or help the poor in facing their limits; and (3) unequal, differentiating units, linked in their economic, social, and political dynamics (such that limits are primarily social and only sometimes biophysically conditioned).

We concentrate here on the contrast between uniform and differentiated analyses, because stratified accounts, while acknowledging the existence of rich and poor, often do not provide an account of the dynamics of formation and maintenance of inequality. Without such dynamics, stratified accounts occupy an uncertain middle ground.[7] Are the policies and other social or technical practices proposed for the poor any different from those from a uniform analysis? If so, more needs to be said—how and why are the proposals supposed to work? If not, then this essay's critique of undifferentiated analyses applies.[8]

In criticizing uniform and aggregate analyses, we must also make clear that, for us, the contrast to global is not local. The local can easily be viewed as a place to become marginalized with respect to more fundamental global trends or, at best, as a mere instance of those trends. Instead, we advocate differentiated analyses that are "locally centered" and "trans-local." That is, one should begin from local situations to keep always in sight the concrete, interconnected social, economic, and environmental dynamics. These will always involve inequality and ongoing differentiation among agents. Furthermore, understanding these dynamics will require tracing their trans-local, -regional, and -national linkages (Arizpe et al. 1994). After all, to continue the scenario above, the land distribution of country B had a history, and probably resulted from land being taken to produce for export, often by foreign or transnational corporations. Understanding locally centered situations and appreciating how they are concretely interlinked is a task of much greater complexity than global analysis, or any account of processes using aggregate and undifferentiated categories. Moreover, the most appropriate point(s) of intervention or engagement are not at all clear in advance of examining the particulars of the situation and the resources one would bring into it. Although this complexity is daunting, the work needs to be done; "think globally, act locally" does not suffice.

The contrast between global and locally centered/trans-local is not an issue of simplification for the sake of generality versus accumulation of detail, synthesis versus focusing on particular cases, first approximations

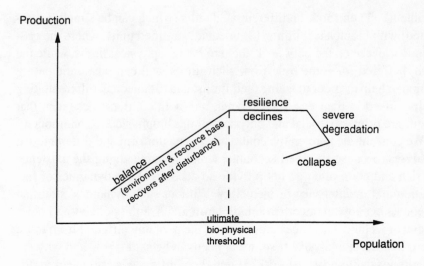

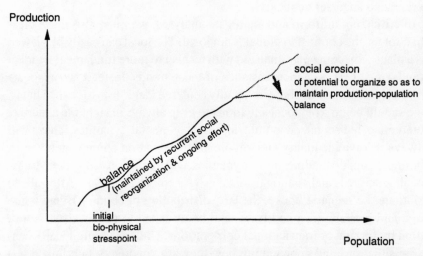

Figure 1. (**upper portion**): Population, production, and environment, showing the sequence of stages implied from global or undifferentiated, aggregate trends; (**lower portion**): the sequence of stages observable in any locally centered situation. (Any aggregate trend is actually an integration of diverse locally centered situations.)

versus more qualified accounts, or choice of temporal and spatial scale. Locally centered/trans-local analysis entails a qualitative change in perspective. Let us illustrate this using some schematic diagrams of the relation between Population growth, production, and environmental degradation (Figure 1).

The upper graph of Figure 1 corresponds to the global formulation of the Population problem. It shows production increasing as Population increases until some ultimate biophysical threshold is reached. Above this, the resource base and the environment begin to become stressed and production cannot keep up with population growth. Eventually, environmental and resource degradation reduces the absolute production capacity, and a Population collapse may occur. Although technological progress may shift the biophysical threshold to the right, it cannot do this indefinitely, for it is natural to reach a production plateau: According to Vitousek et al. (1986), humans in 1986 were consuming 40 percent of the earth's net primary production, so the 100 percent threshold cannot be too far off at present Population growth rates.

In the lower graph of Figure 1, biophysically contingent stress points have already been reached in various places at various times, and the local peoples have initiated processes of reorganizing social institutions (in the broad sense of the term) and technology so that production could keep pace. (In fact, the world's Population would never have been able to reach the level of 40 percent net primary production consumption without such processes). At any Population size, the balance could be upset if the bases for these institutions and for the use of technology were undermined. When this process falters and environmental/resource stress or degradation occurs, there are always social forces (analyzed in the next section) to account for the erosion. (The forces integrate both local and external changes—that is, are locally centered/trans-local.) Conversely, given that people work with and modify institutions and technology in seeking to respond to crises, they will have the most chance of recovering some balance if they appreciate the social character of the crises they have been and will be confronting. Although a society's demands on resources and the speed of growth in those demands will condition the possibilities and processes of socio-technical reorganizing or erosion, the sheer size of the Population or its resource demands do not, either by themselves or as some "root causes," determine the timing and nature of the environmental degradation.

Two different notions of balance are represented in the contrasting formulations (Figure 1, upper versus lower). The conventional view of the Population-resource use system is that the different forces stimulating the Population to exploit its resources push the system out of its basic condition of balance to which it will return if the forces diminish. This might be pictured as a ball in a basin (which becomes shallower as one moves through the stages of the upper graph in Figure 1.) If, however, the forces push the system over a threshold, one considers the resource to be overexploited. Once outside the basin, the system rolls down the hill to a new stability condition, usually the resource's degradation or extinction. A contrasting picture is that in many places, the environment (e.g., topsoils, rainforests, water bodies, and so on) has already been deeply transformed and, thus, a local stress point has long been reached and surpassed, but various social conservative forces are sustaining the resource from rolling down the hill (which may become steeper as one moves through the stages of the lower graph in Figure 1) into a situation of degradation. Therefore, it is the failure of these forces to work efficiently which may precipitate the resource system falling into degradation and extinction.

The second view has several further implications for a critique of the Population problem, which we list here and support in the sections to follow:

- Population size and growth are not at the center of the dynamics of social erosion. The abstract dynamics of Population growth do not provide a sufficient description of the causes of environmental and resource degradation (Palloni 1994).[9]
- The local Population figures aggregated into any global or regional Population figure measure only one facet of the locally centered social-economic-environmental dynamics producing such growths. Given this, demography (i.e., the study of Population as a system) is not a natural, sufficient, or powerful framework in which to explain Population growth.[10]
- Regulation of Population growth cannot be achieved independently from changes whereby the poor and less powerful regain some capacity to reorganize their local social institutions and technology, and, through this, some greater control over production and consumption (Ostrom 1990; McCay and Jentoft 1996).

Our introductory mapping is nearly complete; the general coordinates of our position and the vector of our orientation should be clear. But, before moving into detailed arguments, we want to establish our distance from three formulations that are probably well known to environmentalist readers—formulations that, like ours, also point to the potential for reorganizing social institutions and technology so that production and Population keep pace with each other.

The anti-Malthusian Julian Simon celebrates the power of creative individuals who, when unfettered by government restriction and uninhibited by neo-Malthusian pessimism, are able to generate the knowledge, inventions, and other responses needed to forestall resource scarcity (Simon 1990). The implied account of how institutions and technologies that enhance production are generated (and undermined) is simplistic, based primarily on his free enterprise, anti-government ideology. Like those he seeks to debunk, Simon's analyses are abstract and statistical; not surprisingly, his hyperoptimism discounts the extent to which locally centered/trans-local crises are already widespread and require attention.

Large-scale international aid efforts, in contrast, begin from the position that most local and national institutions in poor societies are inadequate to keep production and Population in line. This "institutional insufficiency" is used to justify the focus of aid being placed on modernizing the technology or the institutions, or otherwise adjusting the structure of the economy (Southgate and Basterrechea 1992). Such a focus, by discounting the potential for endogenously generated reorganization, undermines one of the bases we hold to be essential for generating sustainable institutions of production, reproduction, and consumption.

Finally, we are sympathetic politically and ethically with those who call for empowerment of the poor, of local communities, or of women, and who insist that this empowerment must be part of efforts to alleviate poverty or improve reproductive health and choice (Cohen 1993; Dixon-Mueller 1993; Population Reference Bureau 1993; Institute for Philosophy and Public Policy 1993; Stein 1995). Nevertheless, our emphasis differs. We do not want any empowerment efforts to be justified in terms of their effectiveness in reducing Population growth (see note 4).

Population does not need to be the focal (independent or dependent) variable in any analysis of causes or formulation of responses to economic hardship and environmental degradation. Instead, it is important to understand the processes by which the capacity of women and the poor have become unable to respond effectively to economic, environmental, and other social changes, and to make this understanding central in designing interventions to reverse these processes. We develop this argument in the following section.

LABOR SURPLUS AND INSTITUTIONAL INSUFFICIENCY: CONTRASTING ANALYSES OF THE SOCIAL DYNAMICS OF POOR POPULATIONS AND ENVIRONMENTAL DEGRADATION

Neo-Malthusian environmentalism has an implied view of the relationships among population, labor supply, the social organization of production, technology, and environmental degradation. Let us examine this by contrasting it with other views of the origins and nature of institutional insufficiency common in poor societies.

Boserup and Lewis

As a starting point for our discussion of the social-economic-environmental dynamics of human populations, let us consider Esther Boserup's (1965) still influential argument. She challenged the conventional neo-Malthusian position that population growth must outstrip resources, arguing that population, resources, and technology are linked in a progressive manner, in which population pressure provides a useful economic stimulus to technical and institutional innovation. In particular, population pressure on land stimulates agricultural progress and institutional adaptation, which then allows unprecedented levels of population concentration. In the light of the large-scale historical evidence on the evolution of the world's agriculture, Boserup's argument seemed quite reasonable. However, Boserup's mechanism cannot be operating universally, since technological and institutional adaptation is not now occurring in most poor societies. Why has the Boserupian stimulus stopped functioning, allowing overpopulation or, for that matter, allowing resource depletion of any kind?

One angle to approach this question is through exploring how market mechanisms fail to provide an efficient and flexible monitoring system of natural resource scarcity. Thus, any lag in transmitting information of resource depletion due to population increase (or any other cause) results in insufficient technological and institutional responses. (We say something about this approach later.) A second angle, the one we work from here, derives from the idea that most of the poor human populations of the world constitute an unproductive, capital-scarce, and otherwise institutionally insufficient "labor surplus."

The concept of labor surplus was originally introduced by Arthur Lewis (1954) to characterize what he considered the primitive productive condition of "traditional" societies in underdeveloped countries. According to

Lewis, in those societies the physical relationship between a large population and scarce resources led to a null marginal productivity of labor. Supposedly, the existence of a large portion of nonproductive labor, or surplus labor, in the premodern sector provided developing countries with a mechanism of growth that was economically (and hence institutionally) neutral to rural productivity. As Lewis pointed out, in such conditions "the holding...is so small that if some members...obtained other employment, the remaining members could cultivate the holding just as well" (Lewis 1954, p. 141). Such labor surplus may also be conceived, more fundamentally in our view, as labor that does not endogeneously reorganize its institutions and use of technology to improve its production efficiency. When poor, overpopulated human societies constitute a labor surplus with such a restricted ability to reorganize locally, their own increasing numbers and demand do not stimulate agricultural progress and institutional adaptation. With this insight, one can begin to see why Boserup's mechanism is by no means universal.

The origins and dynamics of the poor's inability to reorganize collectively in response to new challenges, and thus of this "reorganizing-restricted" labor surplus, may be subject to different explanations. In his theory of development, Lewis implicitly assumes it is due to intrinsic characteristics of large traditional populations (their social and economic institutions are primitive, weak, and inefficient) and is determined by physical resource restrictions (physical capital and land are scace relative to population size). Given the reorganizing-restricted character of this rural labor surplus, the only way to transform it into productive labor is through its absorption by another sector, namely, a modern industrialized sector. The surplus, in short, must be upgraded as human capital.

Neo-Malthusian Institutional Insufficiency

Lewis' interpretation of rural labor surplus underlies most analyses of the relation between poor rural societies and ecological change, including neo-Malthusian ones. Unlike Lewis, however, neo-Malthusians have little confidence in industrialized sectors to absorb the labor surplus, which leads them to maintain an emphasis on the situation of the poor. In particular, poor human populations are held to be, as a consequence of their institutional insufficiency, deeply involved in three vicious circles:

1. The poor mismanage or deplete their resources, which in turn reduces land productivity, increases environmental degradation, and limits future income options.

2. Because of their lack of physical and human capital and the distortions in the markets, prices, and credit systems in which they operate, the poor are inefficient and uncompetitive producers. This further restricts their capacity to acquire necessary new capital and overcome their economic disadvantages. (The first circle is emphasized more by environmentalists; the second by international financial institutions such as the World Bank.)

3. A central feature of the resource mismanagement by the poor (in circle 1), enhanced by their lack of economic security (in circle 2), is that they are not able or willing to regulate their numbers—which, on average, leads to further impoverishment.

Given these vicious circles connecting impoverishment, environmental resource degradation, and population growth, neo-Malthusians can conceptualize poor populations' dynamics in terms of their increasing rates of consumption and, hence, resource depletion and rapid approach to biophysical limits. The policy prescriptions that follow are directly related to stopping the population/labor surplus from increasing and thus depleting the natural resources. This view is moderated by a small concession to Lewis' idea of upgrading human capital, namely, that reproductive education, health programs, and direct welfare assistance may help break the vicious circles between poverty, population, and environmental degradation and gradually transform the rural poor into a sustainable sector. (The power of such programs has not, however, been demonstrated in any practical way; Pritchett 1994.)

In spite of its great popularity, Lewis' assumption on the origins and nature of labor surplus is weakly supported by historical and contemporary studies. Moreover, anthropological, sociological, and historical studies on the transformation of precapitalist poor societies under the impact of new capitalist social relations seem to point toward explanations quite different from any intrinsic weakness and inefficiency of the poor's social and economic institutions. In the following two subsections, we present an explicit and more powerful interpretation of the causes and dynamics of any labor surplus that is restricted in its ability to reorganize locally. Through this framework, one can better understand the biophysically conditioned social limits experienced by the poor.

A Structural and Social Psychological Account
of Institutional Insufficiency

Poor populations of the capitalist world are not economically autonomous but participate in a complex arrangement of institutional and economic relations with other social groups and the state, which involve
market and nonmarket transactions at the local and regional levels. In this
context, several structural factors and policies contribute to the ongoing
production of generalized poverty and the disruption of social organization. The factors documented—by authors such as Bartra (1979), Bhaduri
(1983), Bardhan (1984), Binswager and Rosenzweig (1986), Cornia et al.
(1987), De Janvry and García Barrios (1988), García-Barrios and García-
Barrios (1990), and Watts and Peet (1993)—include:

- Unfavorable economic policies and public investment priorities
  (especially with the onset of the debt crisis in the 1980s);
- Structural and institutional contexts that are unfavorable to rural
  development, including inegalitarian land tenure systems and institutional biases against smallholders in the definition of public goods
  and services and in their access to them;
- Economic policies and technological biases that reduce employment
  creation in both the nonagricultural sector and in commercial agriculture;
- Household-specific market failures, economic discrimination, and
  adverse selection in the labor, product, and credit markets;
- Monopolistic power in local formal and informal markets;
- Compulsory transactions which, like usury, lead to the expropriation
  of the poor's resources; and
- Direct private and state coercive violence.

These factors accelerate the macroeconomic disarticulation of agriculture in poor societies. Disarticulation refers to an economic situation in
which investment (in this case, agricultural) is directed toward producing
for export—and so, economic growth depends very little on the development of the internal consumption and production of local communities.
Growth occurs despite—in fact, because of—weak local economies and
low wages (De Janvry 1980). Moreover, many of the transactions entered
into by the poor who face this unfavorable economic context constitute
part of their survival strategies. Once established, however, most of these

transactions become involuntary and compulsory, and many reproduce at the same or greater scale the conditions for poverty and dependency.

These structural conditions constitute a systematic discrimination against the rural poor, which is evidenced in their low productive capacity, the increased instability and uncertainty of their market transactions (i.e., usually in the labor, credit, and product markets), and a reduction of opportunities to establish and maintain viable and stable nonmarket transactions (including the relationships with the state) that could enhance fair local cooperation and circumvent market failures. In short, the structural conditions generate institutional insufficiency, high and increasing transaction costs, and continuing impoverishment. The structural conditions also reduce systemically the capacity of the poor to reorganize endogenously in the face of new challenges, that is, to build up or alter contracts and associations to sustain desirable efficient production, resource management, and technological change. Their institutional insufficiency is reorganizing-restricted.

As a consequence of this reorganizing-restricted institutional insufficiency, there is a labor surplus. Most rural populations must become semi-proletarian, that is, they must survive through off-farm activities involving market transactions. This increasingly demands a high mobility and detachment from the land and social community. This peasant brain and labor drain is not just a matter of external, structural conditions disrupting a community; it interacts with the prevailing social psychology in many serious ways. The individual's decision to incorporate into the market and mobilize the household labor force may be rational from her or his point of view because it increases and stabilizes monetary income. Nevertheless, it acts against the community's institutional arrangements by eroding the bases of local cooperation and social norms.

High population mobility, for example, undermines the blood and ritual kinship systems and reduces face-to-face interaction and the probability of future and repeated relationships between rural agents. Moreover, the manipulation and exploitation of rural people by local patrons and by the state diminishes good will and increases moral hazard, which has led family units to rely on monetary remittances and to abandon cooperation with other units, either in production or consumption, as much as possible. It is also characteristic of populations with high mobility that the significance, prestige, and efficacy of civil and religious systems of authority have been eroded through professionalization and bureaucratization, or through concentration of poorly remunerated official positions in the hands of a few individuals. Less effort is made by authorities to resolve justly the conflicts arising when agents attempt to cooperate; this further promotes the atomization of rural agents and family units. Finally, high mobility and indiscriminate exposure to modern cultural patterns undermine the—generally, already weak—coherence of moral and social structure and language in local communities (e.g., the prestige system, the system of production of native crafts, and the moral economy of reciprocal obligations) that had previously maintained the capacity of mutual identification or trust between the economic agents.

All over the world, peasants and indigenous people organized in "traditional communities" are able only at a very high cost to express their own values, interests, ideologies, norms, and preferences. These values are usually very different from Western ones. Rural and semi-proletarian people in many countries originate from an entire separate history of civilization. The impact of the progressive loss of these values on the social character of the indigenous societies has usually being devastating. While modern societies promise to provide indigenous societies with a wider identity, national elites fail to provide the economic and social support and stimulus necessary to cultivate the crucial components of a new solid social and moral identity. In the face of this, most indigenous societies suffer from socio-pathological depression, a complete loss of communitarian self-esteem, and a lack of confidence in their capacity to govern themselves. This has provided the perfect social psychologial environment for the emergence of strong manipulative governments and oligopolist market structures in rural areas. Even though important remnants do exist in some ethnic groups, the mechanisms to create internal consensus, to negotiate with external private agents or the government, and to mobilize collective labor have weakened or have fallen into disuse.

Recalling our introductory comments, these processes of social psychological disruption and atomization clearly involve differentiating populations and trans-local dynamics. Let us give an example: One of us (García-Barrios) has traced severe soil erosion in a mountainous agricultural region of Oaxaca, Mexico, to the undermining of traditional political authority after the Mexican revolution. Collective institutions had maintained terraces and stabilized the soil dynamics, reducing erosion and maybe even stimulating soil accumulation. This type of landscape transformation also needed continuous and proper maintenance, since it introduced the potential for severe slope instability. The collective institutions revolved around the rich Indian leaders, or "caciques," being able to mobilize peasant labor for key activities. The caciques benefitted from what was produced but were expected to look after the peasants in hard times (a moral economy). Given that peasants felt security in proportion to the wealth and prestige of their cacique, and given the prestige attached directly to one's role in the collective labor, the labor tended to be very efficient. The revolution, however, ruptured this moral economy; transactions and prestige became monetarized following migration to industrial areas and the semi-proletarianization of the rural population; and the collective institutions collapsed (García-Barrios and García-Barrios 1990; García-Barrios et al. 1991).

Implications for Resource Management, Institutional Insufficiency, and Population

The social psychological disruption, together with the effective decrease of the household size due to the migration of the youngest and sometimes most productive members of the family and the continuing poverty of peasant households, reduces the labor and other resources available for land and resource management. That is, the bases not only of economic production, but of environmental/resource conservation and restoration, are eroded with reorganizing-restricted institutional insufficiency. Recall (from the Introduction) the second picture of how a balance between Population and production is maintained or eroded. Rural populations have traditionally stimulated the regeneration of their natural resources, but this depended on collective practices, whose organizational basis is being undermined. Without this basis, externalities can accumulate (such as the production of waste), the environmental carrying capacity decreases, and the biophysically contingent stress points for resource management and economic development are rapidly reached.[11]

The breakdown of local cooperative institutions concerned with the terracing of mountainous areas, evident in Oaxaca, has been more widespread. After the conquest and colonization by European people of many mountainous areas in Latin America, Africa, and the Middle and Far East, the local societies proved unable to maintain such cooperative institutions and agricultural infrastructure rapidly degraded. This history was somewhat repeated after World War II when, due to massive emigration and semi-proletarianization of their inhabitants, societies all around the world, including in South Europe, failed to provide the necessary labor force and cooperation to sustain landscape infrastructure. As a consequence, many terrace systems and agricultural infrastructure are now rapidly degrading, promoting severe soil removal in some areas, downstream siltation, and increasing agricultural poverty.

The degradation of pasture lands in recently colonized tropical areas, in contrast, seems to reflect the inability even to develop (as against sustain) local cooperative institutions of resource management. Weed proliferation is the main cause of the short lifespan of pasture lands (five years or less) and arrested rainforest regeneration in the Amazon (Hecht 1988). In Brazil, an important purpose of both small and large ranches is speculation. Little emphasis is paid to their appropriate management for optimal long-lasting production. Such ill-managed pasture lands become sources of weeds propagules. This increases the probability that neighbouring lands will become infested, even when these are adequately managed (De Janvry and García-Barrios 1988).

Let us now insert this (re)organizing-restricted institutional insufficiency into our picture of Population and environmental degradation. First note that, in the breakdown of terracing and subsequent soil erosion in Oaxaca (and other places), the environmental/resource degradation is linked to an absolute Population reduction in rural areas where peasants are subject to outmigration due to extensive semi-proletarianization or market integration. As populations are greatly reduced, institutional insufficiency may worsen and land abandonment becomes more widespread, in the long run producing the collapse of the carrying capacity of the environment. Therefore, one might turn Boserup's claim about Population increase and technological innovation completely upside down in order to analyze the dynamics of many "modernized" rural societies: Communities with a rapidly decreasing population due to semi-proletarianization may suffer from institutional, technological, and resource degradation because of their inability to rapidly adjust their economic and social institutions to the new circumstances.

Clearly, poverty induction and institutional insufficiency may also occur where rural populations are increasing, as occurs in recently colonized tropical frontier regions where institutions regulating open access are systematically opposed by local interests. The size and/or growth of the Population, however, may not tell us much; instead, in order to explain the escalation of consumption pressures on land, one needs to examine the structural conditions of land tenure and resource distribution, and larger socioeconomic forces that restrict employment creation and enhance social and geographical mobility. These pressures may be occurring even where ultimate carrying capacity is far from being reached; this situation can be seen, for example, in most rural and forested areas of Latin America (Collins 1987) and Africa (Little 1987; see also Arizpe et al. 1994).

Similarly, environmental problems associated with expansion and intensification of agriculture, such as overdrawn aquifers and polluted runoff, occur in countries with high Population densities, such as India. Yet, the problems also occur in countries with no absolute or large consumption pressures, showing that a neo-Malthusian emphasis in responding to environmental and agricultural crises is misplaced. In fact, we are now in a position to comment on the limitations of most Population policies.

The Limitations of Population Policies

The discussion of the structural and social psychological basis of institutional insufficiency shows that a change in the poor's capacity to reorganize their own means of existence is a necessary, but overlooked, condition for attaining sustainability in resource use (Ostrom 1990; McCay and Jentoft 1996). The links between poverty and resource degradation may only be broken by improving the endogenous capacity of poor societies to reorganize and improve their institutional means for collective action and technological change, that is, by improving their capacity to reduce reorganization-restricted labor surplus.

In several ways, neo-Malthusian programs centered on education, development, or welfare assistance are limited by their irrelevance to, and sometimes their erosion of, the capacity of poor societies to reorganize and improve their institutions. How so?

For a start, the invocation of ultimate biophysical limits does not illuminate the current situation that the poor experience. Rather, the causes are to be found in structural dynamics producing and maintaining poverty, which determine the moral and social context in which poor households define their rational responses and survival strategies. Knowledge of these causes enables us to understand why, at times, the poor increase the number of their expected offspring and "mine" natural resources.

Even when education focuses on technological and organizational development, it may be misdirected if the problem is not the absence of local education or culture but the impossibility of the people's using their sometimes profound local environmental knowledge to solve the problems of production and of the environment. In recent years, the overuse or careless use of mechanical technology and agrochemicals has created major ecological threats for the rural and urban populations, spreading doubts about whether modern technologies are really better than traditional ones in the long run. Extensive research has shown the conservation potential of the very sophisticated land management practices embodied in traditional knowledge systems and in modern agroecology (Richards 1983; Hernández-Xolocotzi 1985; Altieri and Anderson 1986; Wilken 1987). This same research, however, also shows the lack of effective use and rapid deterioration of this knowledge basis.

Diffusion among rural societies of any type of labor- or organization-intensive technology, even when this has been developed according to the patterns of local culture (e.g., appropriate technologies), is difficult. Making use of externally supplied education, knowledge, and culture has an opportunity cost to poor populations and, in the absence of a proper institutional framework, may be difficult to transform into useful resources for survival. Rejection of, or resistance to, programs thus is likely. (The same is true of many programs concerned with health and reproductive education.)

Economic, social, and political organizing by the poor is, unlike simple externally supplied education, often threatening to national governments. National governments and international funds usually avoid providing resources (including organizational education) to stimulate such reorganizing. As a consequence, the potential benefits of cultural development are not realized.

There is an emerging consensus that welfare assistance to combat poverty should be changed to become clearly separated from production subsidies, since subsidies produce distortions in the market prices and, hence, generate welfare "basket cases." Since such schemes do not address reorganizing-restricted institutional insufficiency, they can only partly alleviate the poverty/resource degradation vicious circle. Furthermore, many individuals perceive their possibility of participating in production as a basic human need and a right, since they correctly visualize their own possibility of a fulfilled life as a synergistic process in which there is a convergence and feedback of production, economic, moral, religious, social, and political activities. If this is so, the development of their productive capacity and their welfare are all inseparably tied together, and as such, the recognition and respect of their productive values conferred by other people lead to cooperation, reduced resource wastage, incentives for creative work, and reduced expenses for public safety. On the other hand, inequality and injustice give rise to social psychological disorder and inefficiency in the long run.

Finally, given that continuous external assistance degrades the cultural, moral, and psychological basis of individuals and societies, resource degeneration may even be exacerbated.

MORALISTIC AND TECHNOCRATIC
ENVIRONMENTAL DISCOURSE

The examples and interpretations in the previous sections indicate that there are many conceptually and empirically challenging issues that need further investigation in order to develop a sophisticated understanding of the relationships among population, social organization, technology, and environment in different situations. We are well aware, however, that we are not the first to offer a critique of the Population problem.[12] Despite strong criticism, the belief persists that environmental concerns necessitate first and foremost Population (and population) control measures.[13] Given that this reductive formulation of socio-environmental change holds a strong attraction for many environmentalists, the sources of its popularity need to be addressed. This should provide additional purchase for influencing people to abandon a neo-Malthusian framework; from our experience, the conceptual and empirical challenges alone are enough to achieve that end.

To develop this line of discussion, we shift to a different style of analysis. Whereas in the previous section we pointed to the conceptual and empirical weaknesses of neo-Malthusian environmentalist accounts of socio-environmental change, now we interpret this area sociologically. The sociology of science has, over the decades, observed the shaping of what counts as scientific knowledge, especially during controversies, and concluded that the truth of any contested result is rarely sufficient to account for its acceptance; and, conversely, falsity is rarely sufficient to account for its rejection (see, e.g., Collins and Restivo 1983; Star 1988; Woolgar 1988). The previous two sections have indicated that other analyses of the dynamics of population and resources exist. Therefore, the fact of the exponential growth of the global population and of many regional populations is not sufficient to account for why people believe in the Population problem. Instead, we suggest that one can gain critical perspective on adherence to the idea of overpopulation by way of four propositions, adapted from Taylor and Buttel's (1992) discussion of global environmental discourse, which we state and then develop:

1. It is fairly obvious that most environmental analyses are performed for some sponsor or client, or at least with some agency that would implement policy in mind. What is not so obvious is that certain courses of action are facilitated over others in the very formulation of scientific knowledge—in the problems chosen, categories adopted, relationships investigated, and degree of confirming evidence required (Taylor 1989, 1992, 1995). Politics—in the sense of courses of social action pursued or favored—are not merely stimulated by scientific findings; politics are woven into the fabric of scientific knowledge. Science-as-politics warrants an interpretive style of analysis, in which one pays attention to the language and rhetoric of any discourse, and no longer takes literally what is said.

2. In the Population discourse and, more generally, in steady-state discourse, two allied views of politics—the moral (in the sense of appealing to individuals to join a cause) and the technocratic—have been rhetorically privileged. Both views of social action emphasize people's common interests in controlling growth. Inequalities among nations and among social groups within nations are obvious to all who speak about Population problems. Nevertheless, the Population formulation steers attention away from the social, political, and economic dynamics behind people's unequal responsibilities for causing and alleviating environmental problems. People know that there is a Population Problem, in part because they act as if they are unitary and not severally differentiated "we's."[14]

3. Inattention to the localized social and economic dynamics involving population change ensures that scientists, environmentalists, and policymakers are continually surprised by unintended outcomes, unpredicted conflicts, and undesired coalitions.

4. To the extent that people attempt to focus on over-Population, to stand above such coalitions and the conduct of such conflicts, and to discount their responsibility for the unintended outcomes, they are more likely to facilitate increasingly coercive responses to environmental degradation.

Let us begin our elaboration of these propositions by identifying a contradiction or, at least, a tension in our argument. Acceptance of the first proposition, when combined with the previous sections' emphasis on differentiated analysis, should lead one to seek multifaceted analyses of the politics woven into environmental knowledge, in preference to or before making any generalizations. Yet, clearly, the other three propositions are generalizations. Moreover, the second proposition might, by analogy, lead one to interpret any such generalizations as an attempt to avoid dealing with the particulars, messiness, and other difficulties of achieving change (here, the change to be achieved would be in environmental analysis and policy). We acknowledge the contradiction but do not have room in this essay to describe any differentiated, locally centered, trans-local analyses of the politics of environmental knowledge-making (Taylor 1992, 1995). Nevertheless, we think that the generalizations in this section are provocative and useful heuristics to bring about some much-needed reflection on the politics of knowledge. At the same time, we recognize that our raised level of polemic will not bring everyone around to our side. With this admission of this essay's limitations, let us forge ahead.

Recall the scenario of countries A and B from the Introduction. As we noted earlier, many more considerations are required for a sufficient description of the social dynamics in which people contribute differentially to environmental problems. Nevertheless, the conclusion can be drawn that any demographic analysis separated from differentiating social dynamics is taking a definite political stand. Everyone, of course, acknowledges that there are rich and poor, that the rich consume more per capita, and that it may be poverty that compells the poor when they "mine" their resources. Acknowledging the statistics of inequality does not, however, constitute an analysis of the dynamics of inequality. In the absence of serious intellectual work—conceptual and empirical—heartfelt caveats about the rich and the poor do not substantially alter the politics woven into the neo-Malthusian framework.

The politics of neo-Malthusian discourse can be characterized by allied moral and technocratic tendencies. Moral politics emphasizes that everyone must change (reduce their family size) to avert catastrophe. Coercion is rejected; each individual must make the change needed to preserve the environment. Technocratic, on the other hand, signifies that objective analyses (of population growth) identify the severity of the crisis, and technical measures (e.g., contraception and sterilization) are developed and provided (with the appropriate policy stimuli) for individuals and countries to adopt. There is little tension, however, between voluntary individual responses and the managerial-technical ones. They are alike in attempting to bypass the political terrain in which different groups experience problems differently and act accordingly.[15]

They appeal to common, undifferentiated interests as a corrective to governance that is either, in the case of moral politics, corrupt and self-serving or, in the case of technocratic politics, naive and scientifically ignorant. Moreover, like all appeals to universal interests, special places are implicitly built into the proposed social transformations—the moralist as guide/educator/enlightened leader, and the scientist as analyst/policy advisor (Taylor 1988; Taylor and Buttel 1992). In fact, in the absence of any analysis of differentiated interests, a focus on Population as a system offers logically no other standpoints for an environmentalist to take.

So far, this is an interpretation based on the conceptual structure of neo-Malthusianism Population discourse. That is, the rhetorical privileging of moral and technocratic responses is entailed by the aggregate categories of demography, by the invocation of ultimate limits as against analyses of dynamics of differentiation, and by the focus on technical problems, such as contraceptive delivery, as against social-political reorganizing. We have, however, observed similar conceptual structures and privileging of the moral and technocratic more generally in environmental discourse (Taylor 1988, 1992; Taylor and Buttel 1992). What pragmatic and practical reasons might explain why environmental scientists are susceptible to these moral and technocratic tendencies?

The first possible reason is that moral recruitment to a cause and appeals to universal interests can be effective as political tactics; human rights campaigns in times of severe political repression demonstrate that. More generally, political mobilization usually depends on stressing a commonality of interests and playing down differences. Similarly, a technocratic outlook is an understandable orientation for scientists who would rather apply their special skills as best as they can to benefit society, rather than expend energy in political organizing for which they usually have little experience or aptitude. But, perhaps the most important reason why a scientist might be susceptible to the moral and technocratic tendencies is the language that predominates in global environmental discourse (of which neo-Malthusian environmentalism is just one strand). It seems very difficult for anyone to engage in that discourse and enlist others to his or her point of view without slipping into the languages of moral recruitment and education, or of management. This was brought home to us in reviewing the discussion papers and notes circulated in preparation for a volume on equity and sustainability (Smith et al. 1994) and in reading an editorial for the journal *Conservation Biology* (Meffe et al. 1993). We will quote from these sources to illustrate how language that is familiar and well meaning partakes of these two tendencies; many other texts would, however, have made our point equally well.[16]

In these papers, we read of a call for "a total picture of the world" and "rechannel[ing] activity into sustainable forms," phrases that conjure up the hubris of a technocrat. Moralistic language was, however, more pervasive. Recruitment to the cause of responding to "our" common prospect was implied in the recurrent use of "we," "our culture," "our existence,"" "humanity," and in phrases such as "our built-in limitations of perception," "time available for us to change our ways." One paper discussed whether "society could be changed quickly enough," basing its claims around behavioral characteristics supposedly given to humans by their evolutionary history; that is, all of us are fundamentally alike, being members of the same species. Individual behavior and social dynamics were often expressed in the same undifferentiated terms, with individual metaphors used for social ideas and without mention of any structure between the individual and society: "Will humankind take the fork leading to disaster or...to survival?" Does society have the "will to alleviate poverty?" "Affluent societies can choose," despite the "perennial foot-dragging of the establishment." "Individuals vary [therefore] societies vary."

The editorial (Meffe et al. 1993) states that conservation biologists "possess the professional responsibility to teach humankind about the perils" (p. 2) of continued Population growth, have "the obligation to provide leadership in addressing the human population problem and developing solutions" (p. 2), and are able to "help promote policies to curb rapid population growth" (p. 3). "The population problem is stunningly clear and ought to be beyond denial" (p. 2). "The human species ignores or denies" the impending calamity (p. 2). (Presumably those who draw attention to the Population problem are excused from this species collectivity.) A brief mention of the "critical importance...of educating and empowering women" (p. 3) in the next to last paragraph hints that all people might not be equally responsible, but the conclusion returns to the dominant undifferentiated formulation: "Action is needed from everyone, at every turn...[in the cause of] human population control. Life itself is at stake" (p. 3).

Once one starts to notice undifferentiated language, it seems to be used everywhere, even by many who would not choose to be labeled technocrat or moralist. The challenge then is to advance this interpretation in a way that shifts people away from moral and technocratic forms of science-politics. It is obvious that we oppose neo-Malthusian environmentalism; we consider its science to be conceptually inadequate and often empirically superficial, and we want to assert the need for a differentiated politics in all environmental discourse. How can we move the discussion of Population and environmental degradation in this direction? Notice that we have pointed to the practical facilitations of the moral and technocratic tendencies, so we do not expect these tendencies to be undermined by a mere counter-interpretation, that is, something working mostly on an intellectual and textual level. The most practically oriented approach, as mentioned earlier, would be to go beyond the generalizations above, to investigate particular cases of environmental knowledge-making and, based on the diverse facilitations observed, to contribute to building conditions favorable to alternative science and politics in that case. Nevertheless, the step we take here, in which we raise our polemical level and push our generalized critique further, might lead neo-Malthusians to build their own ways of changing.

We have argued that certain politics (here, the moral and technocratic tendencies) and the science that facilitates them are not dictated by the nature of reality. We have intended, thereby, to establish that scientists and other social agents choose to contribute to such science-politics. They are, thus, partly and jointly responsible for their consequences. In order then to urge neo-Malthusians (both self-professed and by disposition) to acknowledge that responsibility, let us stress that their science-politics does have consequences. Policies based on abstract aggregated analyses have unintended effects and undesirable surprises are inevitable. When these policies are promoted through crisis rhetoric that feeds on fears about the future, coercion and violence become more likely.

For example, in the early 1980s in Chiapas in southern Mexico, villagers became angry when they discovered that internationally funded health workers were sterilizing women after childbirth without their consent. The villagers killed two of these workers, only to have the government call in the military to raze the village in retaliation. This may be an extreme case, but it is not merely "unfortunate." The underlying conceptualization of Population control policies is implicated in the causes of such cases in the following way: The Population Problem translates readily into medical and clinical measures to reduce birth rates, which do not seem to require analysis of particular social and economic dynamics. In the absence of such analysis, there is a much reduced chance that resistance will be anticipated, understood, or tolerated by the international agency and the government. Also, the belief that institutions in poor societies are generally weak and corrupt excuses the heavy-handed action by some states, without shedding light on why some poor states are not so heavy-handed. Moreover, the Chiapas event is not an isolated case. In India during the 1960s and 1970s, especially during the emergency of 1975-1976, population programs resulted in injuries and deaths (Blaikie 1985, p. 98ff.). In the resistance and revolt that occurred, democratic aspirations were linked with opposition to family control programs; this is surely an unfortunate coalition in the eyes of most Western environmentalists.

Over the last generation, Population growth has declined in many countries and, in some cases, statistically significant effects of Population control programs have been discovered (but see Pritchett 1994). Yet, the successful programs have piggy-backed upon other social changes favoring reductions in birth rates, such as the employment of women in the formal work force, reductions in infant and child mortality, the increased value of educating children at the same time that this education is incurring a cost to the family, and so on (Blaikie 1985). Analysis of the differentiating social and economic dynamics of particular situations would not only help explain the occasional successes but also help to plan the broader family welfare programs needed to accompany birth control programs. Conversely, such analysis would help in anticipating the ways that the broader measures, such as adult literacy campaigns or the development of appropriate technology, can be undermined by the dynamics of labor scarcity or by those whose interests are threatened in some way. For these reasons alone, one might abandon the Population problem as a framework for analysis and action. But let us push the critique yet one step further: The violent and coercive dimensions of the Chiapas program, also associated with programs in India in the 1960s and 1970s, warrant our examining the Population framework for any inherent tendencies to coercion or violence.

The moral posture of most environmentalists—lifeboat ethicists (Hardin 1972) and certain biocentric deep ecologists (see Bradford 1987) aside— is to support sustainable, liveable, and equitable futures for all people, free from economic and political coercion. In fact, many neo-Malthusian environmentalists reinforce their appeal for population control on the grounds that without it coercive measures will surely be taken when the crisis becomes more severe (see Ehrlich and Holdren 1971, p. 1216). The Population framework, however, works against this professed commitment in many ways, which we summarize now.

The use of undifferentiated categories—such as population, affluent societies, and human nature, as described above—facilitate moral and technocratic discourse that provides little purchase either in explaining the outcome of Population control programs or in generating successful ones.

The lack of analysis of the interrelations among population, social organization, technological change, and the environment makes any analysis of the interrelation between the affluent and poor difficult and, at best, holistic and simplistic. This, in turn, facilitates the abstraction of considering the poor and the affluent separately—in fact, as essentially different types in their institutions, consciousness, and social possibilities (Ehrlich and Holdren 1971; Cohen 1995).

The essentialistic conception of affluent and poor people permits a simplistic analysis of the possibilities of productive and creative institutional responses in societies that may be classified, on average, as affluent or poor. Furthermore, it reinforces the moral authority that accrues to the affluent by virtue of their potential to educate or otherwise intervene through education, and of capable political and technical institutions to respond to environmental problems.

Several factors combine to make the discourse and practice of neo-Malthusian environmentalism and Population control susceptible to shifting into a coercive posture: frustration in the face of failed Population control programs, the urgency of the environmentalists' crisis rhetoric (e.g., Ehrlich and Ehrlich 1990; Meffe et al. 1993), the lack of any differentiated categories and intermediate standpoints between the individual and society, the essentialistic contrast between capable and fair institutions in affluent societies and weak and corrupt ones in poor societies, and the moral authority to intervene. Coercion is not just an abstract possibility but one environmentalists more generally must pay attention to, as Nancy Peluso's (1993) analysis of the coercive dimensions of internationally endorsed conservation schemes, such as wildlife reserves in Kenya and forest conservation in Indonesia, indicates. Many conservation schemes require or assume state control over natural resources, whereas this is often resisted by local peoples who have been gaining some of their livelihood from the resources in question—elephant tusks, game, products from the forests, and so on. Conservation schemes have thus given the state and militarized institutions opportunities to gain more control over territory and peoples under a seemingly benevolent banner.

A different path to coercion derives, ironically, from the endorsement by various Population theorists and steady-state advocates of the market as a means to protect and promote individual freedom. Contrary to the ideology that market relations are a natural form of interaction among individuals, real markets always have to be constructed. The motivation to construct them generally depends on institutional arrangements that ensure the possibility of accumulation (Rees 1992). Deregulation and the dismantling of the centralized state enhance the power of corporations to dictate more freely the terms of their exchanges. As Marginson (1988) observed, only capital is set free by the free market; people are not. More than a decade of deregulation has enhanced the freedom of corporations to decide the form and location of their investments (Leyshon 1992). Given this, many environmentalists critical of the results of current economic development have made tactical alliances with corporate-led economic policymaking to achieve their aims (Donahue 1990). That is, they have acceded to the power of corporations to control labor and other resources, preferably not in the environmentalists' backyard, but nevertheless, somewhere (but see Daly and Goodland 1994).

CONCLUSION

We have argued that there are many conceptual and empirical reasons to break open neo-Malthusian environmentalist discourse into a social analysis of environmental change (Taylor and García Barrios 1995), to examine the complex ways social organization intervenes between population change and resource use. We have complemented this argument by interpreting the science as rhetoric that favors certain courses of social action. With the aim of challenging concerned Population scientists and neo-Malthusian environmentalists to examine the standpoint they take in research and action, we have argued that a commitment to non-coercion and anti-violence should lead one to avoid moral and technocratic discourse, to dig deeper than the conventional analyses which—in their structure, if not always explicitly—hold poor populations to be the most important drag against the construction of a sustainable world. Neither strongly expressed sympathies for the poor nor reduced personal consumption and fecundity exempt a neo-Malthusian environmentalist from our critique. The complex politics of differentiating local and trans-national resource management and environmental protection require a more strident stance. To prepare to resist any repressive measures undertaken in the name of sustainability requires both serious conceptual and empirical work and difficult political engagement. These challenges are, we think, worthy of the attention of all environmentalists and human ecologists who may want to build a framework for sustainability and equity.

ACKNOWLEDGMENTS

An extended version of this essay will appear as Taylor and García-Barrios (forthcoming). The sometimes strident, even polemical, style of this work reflects its origin in 1992 as Raul García-Barrios's critical contribution to a Pugwash project, "Toward a Sustainable, Liveable and Equitable World" (see Smith et al. 1994). Although we have not addressed directly all of the wealth of literature on population-environment relations that has appeared during the last five years, we have indicated in footnotes what we consider to be key entry points (see, in particular, Arizpe et al. 1994). We thank Chris Finlayson and Reem Saffouri for research assistance, and Elena Alvarez-Buylla, Lee Freese, David Mayer, and Henry Shue for their helpful comments. Travel funds from the Cornell International Institute for Food Agriculture and Development facilitated our collaboration.

NOTES

1. There is a substantial body of research in the social analysis of environmental change upon which the construction of sustainability should be built, but to date this has scarcely taken place. For reviews, see Richards (1983), Watts and Peet (1993), Neumann and Schroeder (1995), and Taylor and García-Barrios (1995). For analyses sympathetic to the position taken in this essay, see Stonich (1989) and Arizpe et al. (1994).

2. The focus on the poor is justified by our observation that the current population-environment discourse, especially that around the formulation of policy, is well developed only where it focuses on the poor (e.g., United Nations Population Fund 1991). Notwithstanding this specific focus, we hope that readers, even those who distance themselves from neo-Malthusianism, will think about how our points can be translated and extended to other areas. In particular, the social-economic-environmental dynamics of the urban poor and of affluent consumers invite similar treatments.

3. Jolly (1994) distinguishes four macroeconomic theories relating population and the environment, and the United Nations Population Fund (1991) provides a clear example of the macroeconomic orientation. A health and choice emphasis for a reproductive practices orientation (Stein 1995) became prominent in the policy arena during the build-up to the United Nations Conference on Population and Development held in Cairo in September, 1994 (see International Women's Health Coalition 1993) and was highlighted in the "Program of Action" endorsed by the conference. Neo-Malthusian environmentalists have been acknowledging this theme in their recent statements about Population control (e.g., Ehrlich and Ehrlich 1990, p. 216; Meffe et al. 1993).

4. It would follow from our analysis of dynamics (in the section of the essay entitled "Labor Surplus..."), if we declared that there were no global problems and asserted that the possibility of future large-scale transitions—involving, say, complex non-linearities in ocean and atmospheric circulation—do not warrant conceptualizing current problems as global problems. However, an interpretation of science as rhetoric requires us to ask who sees problems as global (Taylor and Buttel 1992). In this spirit, we interpret globalized discourse in terms of the particular social actions and politics privileged by it.

5.   Aggregate figures can draw attention to problems requiring attention or explanation. For example, the changing sex ratios of infants in China following the imposition of the one child per couple policy pointed to increasing female infanticide. The fact that demographic transitions were achieved by Taiwan and South Korea, but not the Philippines, after World War II pointed to the importance of successful land and educational reform (Hartmann 1987). Nevertheless, the explanation or the successful policy response requires going well beyond the aggregate figures.

6.   Given the prominent role of population biologists in Population discourse, we note that their field has, over the last decade, begun to pay more attention to the qualitative differences in predictions based on models that distinguish individuals within a population (in terms of their spatial location or other characteristics), compared to the older style of using aggregate variables to describe a population (Huston et al. 1988). This development had not yet occurred in 1971 when Ehrlich and Holdren formulated their neo-Malthusian position explicitly in terms of a mathematical equation, $I = P \times F$, where I is the negative impact of population, P is the population size, and F is a function denoting the per capita impact. Even now, however, the aggregate equation (with F spelled out as $A \times T$—that is, Affluence $\times$ Impact of Technology used—remains central to the analyses of Ehrlich and his current collaborator (Ehrlich and Ehrlich 1990; Meffe et al. 1993).

7.   For examples of such ambiguity, see Mazur (1994) and United Nations Population Fund (1991). In the former, consumption (i.e., of the rich) is in the title of the collection, but most essays, even when critical, focus on limiting Population growth in poor countries. In the latter, a diagram of "links between demographic and natural resource issues" (p. 13) is consistent with a complex account of interconnected social, economic, and environmental dynamics, but the discussion centers on the Population growth and natural resource degradation of poor countries.

8.   Discussions or even condemnations of the disproportionate resource use of the rich do not negate our criticisms. The key question for this essay is what responses are logically consistent with the causes being identified (Harvey 1974), not whether or not an environmentalist shows awareness of inequality among and within nations. Cohen's (1995) comprehensive text on global human carrying capacity is vulnerable to this line of questioning.

9.   By extension, if the Population problem for affluent societies is cast in terms of overconsumption and its consequences, the same insufficiency is true for the abstract dynamics of consumption growth. Instead, we should examine the inability of the affluent to reorganize social institutions and technology so as to ensure satisfaction without compulsive consumption (Roberts 1979; Max-Neef 1986).

10.   Folbre (1994) makes an analogous and much more thoroughly developed case in her analysis of the social, economic, and technological dynamics involved in the reproduction of labor. Mainstream demographers are now also recognizing that, even to explain such established ideas as the demographic transition, they need to invoke mediating variables and undertake fine-grained analyses (see Handwerker 1986; Simmons 1988, p. 91ff.; Preston 1989, pp. 15-16). The field's rationale remains, however, to explain Population changes and the reproduction and migration patterns associated with those changes. Although our critique is focused on neo-Malthusian environmentalism, the shifts of perspective promoted in this paper can be extended to demography. Some younger demographers are preparing the way. For example, Elliott (1994), seeking to bridge aggregate statistical analysis and case studies, proposes a "Boolean-based comparative method" for analyzing the relationship between population and deforestation. Riley (1995) reviews the

challenges that feminist perspectives raise for demographic questions, methods, and theory, and for policy based upon demography. Folbre (1994) should become an important guide or model for such work. See also Stein (1995) and the references cited in Riley (1995) and in Ginsburg and Rapp (1991).

11. Moreover, this has usually occurred in local and national societies where the local authorities and national government have not been able or willing to induce the innovation and diffusion of technologies for sustainable agriculture adapted to the new labor conditions. Nor have the authorities and the state generated institutions able to provide adequate public goods and the means of internalizing the externalities which have arisen.

12. For critiques of neo-Malthusian environmentalism, see Commoner (1971), Harvey (1974), Schnaiberg (1981), and Arizpe et al. (1994). For critiques of neo-Malthusianism more generally and Population discourse, see Finkle and Crane (1975), Bondestam (1980), Hartmann (1987), Mehler (1989), Duden (1992), Nair (1992) Population Reference Bureau (1993), Greenhalgh (1994), and Stein (1995).

13. We have observed that this is especially true among Americans who came of age in the 1960s, a fact inviting social-historical analysis and interpretation.

14. One could analyze the constructions of Population control that are more purely technocratic, but we have chosen to concentrate on the combined moral and technocratic dimensions of neo-Mathusianism, considering this interpretation to have more relevance to environmentalists.

15. We might also describe this as "consensus-seeking." Among steady-state proposals, the Population problem is the only formulation likely to generate much consensus. The resistance of those with an interest in capital accumulation means that other policies with a general and clear impact on development and sustainability cannot generate political consensus at the national and international levels.

16. In the resulting volume, quotations such as those given in the text were accompanied by much more attention to stratification within and among nations, and the technocratic currents were less apparent. (Copies of the precirculated documents are available from the authors.) Taylor and Buttel (1992) indicate how the languages of moral recruitment and management are equally mixed in Meadows et al. (1972), Clark and Holling (1985), and Clark (1989). See also Brandt (1983) and Ehrlich and Ehrlich (1990). Language does not, however, stand on its own, and the reader should not forget the conceptual argument about the aggregate categories and undifferentiated dynamics entailing moral or technocratic responses.

REFERENCES

Altieri, M., and M. Anderson. 1986. "An Ecological Basis for the Development of Alternative Agricultural Systems for Small Farmers in the Third World." *American Journal of Alternative Agriculture* 1: 30-38.

Arizpe, L., M.P. Stone, and D.C. Major, eds. 1994. *Population and Environment: Rethinking the Debate*. Boulder, CO: Westview.

Bardhan, P.K. 1984. *Land, Labor and Rural Poverty: Essays in Development Economics*. New York: Columbia University Press.

Bartra, A. 1979. *La Explotación del Trabajo Campesino por el Capital (The Exploitation of Farm Labor by Capital)*. Mexico City: Ed. Macehual.

Bhaduri, A. 1983. *The Economic Structure of Backward Agriculture*. New York: Academic.
Binswanger, H., and M. Rosenzweig. 1986. "Behavioral and Material Determinants of Production Relations in Agriculture." *Journal of Development Studies* 22: 503-539.
Blaikie, P. 1985. *The Political Economy of Soil Erosion in Developing Countries*. London: Longman.
Bondestam, L. 1980. "The Political Ideology of Population Control." Pp. 1-38 in *Poverty and Population Control*, edited by L. Bondestam and S. Bergström. New York: Academic.
Bongaarts, J. 1992. "Population Growth and Global Warming." *Population and Development Review* 18: 299-319.
Boserup, E. 1965. *The Conditions of Agricultural Growth: The Economics of Agrarian Change under Population Pressure*. London: Allen and Unwin.
Bradford, G. 1987. "How Deep is Deep Ecology: A Challenge to Radical Environmentalism." *Fifth Estate* 22(3): 3-64.
Brandt, W., ed. 1983. *North-South: A Program for Survival*. London: Pan Books.
Clark, W. 1989. "Managing Planet Earth." *Scientific American* 261: 46-54.
Clark, W., and C. Holling. 1985. "Sustainable Development of the Biosphere: Human Actvities and Global Change." Pp. 474-490 in *Global Change*, edited by T. Malone and J. Roederer. Cambridge: Cambridge University Press.
Cohen, J.E. 1995. *How Many People Can The Earth Support?* New York: Norton.
Cohen, S. 1993. "The Road from Rio to Cairo: Towards a Common Agenda." *International Family Planning Perspectives* 19: 61-66.
Collins, J. 1987. "Labor Scarcity and Ecological Change." Pp. 19-37 in *Lands at Risk in the Third World: Local Level Perspectives*, edited by P. Little, M. Horowitz and A. Nyerges. Boulder, CO: Westview.
Collins, R., and S. Restivo. 1983. "Development, Diversity, and Conflict in the Sociology of Science." *Sociological Quarterly* 24: 185-200.
Commoner, B. 1971. *The Closing Circle*. New York: Knopf.
Cornia G., R. Jolly, and F. Stewart, eds. 1987. *Adjustment with a Human Face: Protecting the Vulnerable and Promoting Growth*. Oxford: Clarendon Press.
Daly, H., and R. Goodland. 1994. "An Ecological-Economic Assessment of Deregulation of International Commerce under GATT." *Population and Environment* 15: 395-427, 477-503.
De Janvry, A. 1980. *The Agrarian Question and Reformism in Latin America*. Baltimore, MD: The Johns Hopkins University Press.
De Janvry, A., and R. García-Barrios. 1988. "Rural Poverty and Environmental Degradation in Latin America: Causes, Effects, and Alternative Solutions." Paper presented at the "International Consultation on Environment, Sustainable Development, and the Role of Small Farmers," conference, International Fund for Agricultural Development, Rome, October 11-13.
Dixon-Mueller, R. 1993. *Population Policy and Women's Rights: Transforming Reproductive Choice*. Westport, CT: Praeger.
Donahue, J. 1990. "Environmental Board Games." *Multinational Monitor* (March): 10-12.
Duden, B. 1992. "Population." Pp. 146-157 in *The Development Dictionary: A Guide to Knowledge as Power*, edited by W. Sachs. New Jersey: Zed Books.
Durham, W. 1979. *Scarcity and Survival in Central America: Ecological Origins of the Soccer War*. Stanford, CA: Stanford University Press.

Ehrlich, P., and A. Ehrlich. 1990. *The Population Explosion.* New York: Simon and Schuster.

Ehrlich, P., and J. Holdren. 1971. "Impact of Population Growth." *Science* 117: 1212-1217.

Elliott, J.R. 1994. "Population and Deforestation in Central America: An Alternative Approach." Paper presented to the American Sociological Association Annual Meetings, Los Angeles, August 5-9.

Finkle, J.L., and B.B. Crane. 1975. "The Politics of Bucharest: Population, Development and the New International Economic Order." *Population Development Review* 1: 87-114.

Folbre, N. 1994. *Who Pays for the Kids? Gender and the Structures of Constraint.* New York: Routledge.

García-Barrios, R., and L. García-Barrios. 1990. "Environmental and Technological Degradation in Peasant Agriculture: A Consequence of Development in Mexico." *World Development* 18: 1569-1585.

García-Barrios, R., L. García-Barrios, and E. Alvarez-Buylla. 1991. *Lagunas: Deterioro Ambiental y Tecnológico en el Campo Semiproletarizado (Lagunas: Environmental and Technological Deterioration in the Semi-protelarioanized Countryside).* Mexico City: El Colegio de Mexico.

Ginsburg, F., and R. Rapp. 1991. "The Politics of Reproduction." *Annual Review of Anthropology* 20: 311-343.

Greenhalgh, S. 1994. "Controlling Births and Bodies in Village China." *American Ethnologist* 21: 3-30.

Hall, C.A.S., R.G. Pontius, L. Coleman, and J-Y. Ko. 1994. "The Environmental Consequences of Having a Baby in the United States." *Population and Environment* 15: 505-524.

Handwerker, W.P. 1986. "Culture and Reproduction: Exploring Micro/macro Linkages." Pp. 1-28 in *Culture and Reproduction: An Anthropological Critique of Demographic Transition Theory,* edited by W.P. Handwerker. Boulder, CO: Westview.

Hardin, G. 1972. *Exploring New Ethics for Survival.* New York: Viking Press.

Hartmann, B. 1987. *Reproductive Rights and Wrongs: The Global Politics of Population Control and Contraceptive Choice.* New York: Harper & Row.

Harvey, D. 1974. "Population, Resources and the Ideology of Science." *Economic Geography* 50: 256-277.

Hecht, S. 1988. "Conversion of Forest to Pasture in Amazonia: Some Environmental and Social Effects." Working paper, Department of Urban and Regional Planning, University of California, Los Angeles.

Hernández-Xolocotzi, E. 1985. *Xolocotzia, Tomo I: Revista de Geografía Agrícola (Xolocotzia, Vol. I: Review of Agrarian Georgraphy).* Mexico City: Universidad Autónoma de Chapingo.

Huston, M., D. DeAngelis, and W. Post. 1988. "From Individuals to Ecosystems: A New Approach to Ecological Theory." *Bioscience* 38: 682-691.

Institute for Philosophy & Public Policy. 1993. "Ethics and Global Population." *Philosophy & Public Policy* 13(4): 1-32.

International Women's Health Coalition. 1993. "Women's Declaration on Population Policies." *Race, Poverty and the Environment* 4(2): 37-38.

Jolly, C.L. 1994. "Four Theories of Population Change and the Environment." *Population and Environment* 16: 61-90.

Lewis, A. 1954. "Development with Unlimited Supply of Labour." *The Manchester School* 22: 139-192.

Leyshon, A. 1992. "The Transformation of Regulatory Order: Regulating the Global Economy and Environment." *Geoforum* 23: 249-268.

Little, P. 1987. "Land Use Conflicts in the Agricultural/Pastoral Borderlands: The Case of Kenya." Pp. 195-212 in *Lands at Risk in the Third World: Local Level Perspectives*, edited by P. Little, M. Horowitz, and A. Nyerges. Boulder, CO: Westview.

Marginson, S. 1988. "The Economically Rational Individual." *Arena* 84: 105-114.

Max-Neef, M. 1986. *Desarrollo a Escala Humana: Una Opción para el Futuro*. Motala: CEPAUR-Fundación Dag Hammarskjold.

Mazur, L.A. 1994. "Beyond the Numbers: An Introduction and Overview." Pp. 1-20 in *Beyond the Numbers: A Reader on Population, Consumption, and the Environment*, edited by L.A. Mazur. Washington, DC: Island Press.

McCay, B., and S. Jentoft. 1996. "Market or Community Failure? Critical Perspectives on Common Property Research." Working Paper, Ecopolicy Center, Rutgers University.

Meadows, D., D. Meadows, J. Randers, and W. Behrens. 1972. *The Limits to Growth*. New York: Universe Books.

Meffe, G.K., A.H. Ehrlich, and D. Ehrenfeld. 1993. "Human Population Control: The Missing Agenda." *Conservation Biology* 7: 1-3.

Mehler, B. 1989. "History of the American Eugenics Society." Unpublished Ph.D. thesis, University of Illinois, Champaign-Urbana.

Nair, S. 1992. "Population Policies and the Ideology of Population Control in India." *Issues in Reproduction and Genetic Engineering* 5(3): 237-252.

Neumann, R., and R. Schroeder, eds. 1995. "Manifest Ecological Destinies: Local Rights and Global Environmental Agendas." *Antipode* 27(4): 321-448.

Ostrom, E. 1990. *Governing the Commons: The Evolution of Institutions for Collective Action*. New York: Cambridge University Press.

Palloni, A. 1994. "The Relation between Population and Deforestation: Methods for Drawing Causal Inferences from Macro and Micro Studies." Pp. 125-165 in *Population and Environment: Rethinking the Debate*, edited by L. Arizpe, M.P. Stone, and D.C. Major. Boulder, CO: Westview.

Peluso, N. 1993. "Coercing Conservation: The Politics of State Resource Control." Pp. 46-70 in *The State and Social Power in Global Environmental Politics*, edited by R. Lipschutz and K. Conca. New York: Columbia University Press.

Pimentel, D., R. Harman, M. Pacenza, J. Pecarsky, and M. Pimentel. 1994. "Natural Resources and an Optimum Human Population." *Population and Environment: A Journal of Interdisciplinary Studies* 15: 347-369.

Population Reference Bureau. 1993. "What Women Want: Women's Concerns about Global Population Issues." Report to the Pew Charitable Trust Global Stewardship Initiative, June 1993.

Preston, S. 1989. "The Social Sciences and the Population Problem." Pp. 1-26 in *Demography as an Interdiscipline*, edited by J. Stycos. New Brunswick, NJ: Transaction.

Pritchett, L. 1994. "Desired Fertility and the Impact of Population Policies." *Population and Development Review* 20(1): 1-55.

Rees, J. 1992. "Markets—The Panacea for Environmental Regulation?" *Geoforum* 23: 383-394.

Richards, P. 1983. "Ecological Change and the Politics of Land Use." *African Studies Review* 26: 1-72.

Riley, N. 1995. "Challenging Demography: Contributions from Feminist Theory." Working paper, Department of Sociology, Bowdoin College.

Roberts, A. 1979. *The Self-Managing Environment.* London: Allison and Busby.

Schnaiberg, A. 1981. "Will Population Slowdowns Yield Resource Conservation? Some Social Demurrers." *Qualitative Sociology* 4(1): 21-33.

Simmons, O. 1988. *Perspectives on Development and Population Growth in the Third World.* New York: Plenum.

Simon, J. 1990. *Population Matters: People, Resources, Environment, and Immigration.* New Brunswick, NJ: Transaction.

Smith, P.B., S.E. Okoye, J. de Wilde, and P. Deshingkar, eds. 1994. *The World at the Cross-roads: Towards a Sustainable, Equitable and Liveable World.* London: Earthscan.

Southgate, D., and M. Basterrechea. 1992. "Population Growth, Public Policy and Resource Degradation: The Case of Guatemala." *Ambio* 21(7): 460-464.

Star, S. 1988. "Introduction: The Sociology of Science and Technology." *Social Problems* 35: 197-205.

Stein, D. 1995. *People Who Count: Population and Politics, Women and Children.* London: Earthscan.

Stonich, S. 1989. "The Dynamics of Social Processes and Environmental Destruction: A Central American Case Study." *Population and Development Review* 15: 269-296.

Taylor, P. 1988. "Technocratic Optimism, H.T. Odum and the Partial Transformation of Ecological Metaphor after World War 2." *Journal of the History of Biology* 21: 213-244.

_____. 1989. "Revising Models and Generating Theory." *Oikos* 54: 121-126.

_____. 1992. "Re/constructing Socio-ecologies: System Dynamics Modeling of Nomadic Pastoralists in Sub-Saharan Africa." Pp. 115-148 in *The Right Tools for the Job: At Work in the Twentieth Century Life Sciences,* edited by A. Clarke and J. Fujimura. Princeton, NJ: Princeton University Press.

_____. 1995. "Building on Construction: An Exploration of Heterogeneous Construction-ism, Using an Analogy from Psychology and a Sketch from Socio-economic Mod-eling." *Perspectives on Science* 3: 66-98.

Taylor, P., and F. Buttel. 1992. "How Do We Know We Have Global Environmental Prob-lems: Science and the Globalization of Environmental Discourse." *Geoforum* 23: 405-416.

Taylor, P., and R. García-Barrios. 1995. "The Social Analysis of Environmental Change: From Systems to Intersecting Processes." *Social Science Information* 34: 5-30.

Taylor, P., and R. García-Barrios. Forthcoming. "The Dynamics of Socio-Environmental Change and the Limits of Neo-Malthusian Environmentalism." In *The Limits to Markets: Equity and the Global Environment,* edited by T. Mount, H. Shue, and M. Dore. Oxford: Blackwell.

United Nations Population Fund. 1991. *Population and Environment: The Challenges Ahead.* New York: United Nations Population Fund.

Vandermeer, J. 1977. "Ecological Determinism." Pp. 108-122 in *Biology as a Social Weapon,* edited by Ann Arbor Science for the People. Minneapolis, MN: Burgess Publishing.

Vitousek, P., P. Ehrlich, A. Ehrlich, and P. Matson. 1986. "Human Appropriation of the Products of Photosynthesis." *Bioscience* 36: 368-373.

Watts, M., and R. Peet, eds. 1993. "Environment and Development" (Special double issue). *Economic Geography* 69(3-4).

Wilken, G. 1987. *Good Farmers: Traditional Agricultural Resource Management in Mexico and Central America.* Berkeley: University of California Press.

Woolgar, S. 1988. *Science: The Very Idea.* London: Tavistock.

LEGAL REFORM AND LOCAL ENVIRONMENTAL MOBILIZATION

Adam S. Weinberg

ABSTRACT

Since the late 1970s, grassroots forms of organizing have played a central role in the growth of the environmental movement. Despite a tremendous amount of social science research into the history, success, and motives behind grassroots organizing, there has been very little discussion about the role of law and legal reform. Law has been a background assumption which is under-theorized. Critical researchers assume that law is a peripheral force unable to create change. Uncritical researchers assume that legal reform is an indicator of positive change. Here, I explore the relationship between legal reform and grassroots participation. In particular, I examine the impact of two laws specifically designed to empower grassroots groups to partici- pate in ecological conflicts. I argue that: (1) law is important to understand because it plays a large role in shaping the experiences of grassroots groups as they seek to participate in environmental conflicts; (2) legal reform is often conceptualized in ways that exacerbate problems of participation; and

Advances in Human Ecology, Volume 6, pages 293-323.
Copyright © 1997 by JAI Press Inc.
All rights of reproduction in any form reserved.
ISBN: 0-7623-0257-7

(3) crafting legal reforms that support grassroots organizations requires new theoretical frameworks for understanding the relationship between law and environmental conflicts.

INTRODUCTION

Laws and Grassroots Organizing

Since the late 1970s, grassroots forms of organizing have played a central role in the growth of the environmental movement. Between 5,000 and 9,000 active grassroots organizations existed in the United States by the end of the 1980s (Harper 1996; Sale 1993; Cylke 1993). Most of the early social science research concentrated on understanding the characteristics of these groups (Brown and Mikelson 1990; Edelstein 1988; Freudenberg 1984). Both survey data and case studies documented the emergence of a grassroots political phenomenon characterized as a loosely structured movement of community-based groups, regional and statewide coalitions, and national organizations united around the mantra of "Think Globally— Act Locally" (Freudenberg and Steinsapir 1992). Groups tended to be small and were comprised of community residents with little political experience, who mobilized after learning of a direct health risk to their community.

Recent research has emphasized the impact of grassroots organizing. The broad consensus among researchers is that grassroots groups face a difficult task when they try to push for ecologically sound uses of local natural resources. Grassroots groups confront powerful producers and state agents who are more resourceful and who purposively raise the stakes of political participation as part of multifaceted strategies to deny groups an opportunity to voice concerns with proposed actions or to criticize current actions (Gould, Schnaiberg, and Weinberg 1996; Cable and Cable 1995; Cable and Bensen 1995; Pring and Canan 1995). Despite the consistency in this literature, there has been tremendous disagreement over the outcome of grassroots organizing. Research focused on specific cases has generally argued that the power differentials between grassroots groups and producers are too skewed and, thus, the effect of grassroots organizing has been minimal. Most groups are overpowered by producers or ignored by state agents. Even successful groups are found to have accepted reduced outcomes (Gould, Schnaiberg, and Weinberg 1996; Gould 1991, 1993; Pellow 1994; Kousis 1993; Cable and Benson 1993). Research directed

more broadly on the grassroots movement has located beneficial outcomes. The grassroots movement is found to have increased public support
for environmental ideas (Dunlap 1992; Dunlap and Mertig 1992), encouraged more favorable framing of environmental issues (Szasz 1994; Simmons and Stark 1993), and awakened the social consciousness of activists
which filters through a local community (Freudenberg 1984).

Largely missing from this debate has been any discussion of law and
legal reform. For example, "law" and "legal reform" are remarkably absent
as categories from the indexes of major books published in the last few
years (Harper 1996; Schnaiberg and Gould 1994), as well as from chapters
in collections of edited essays (Dunlap and Mertig 1992), and paper sessions at professional meetings. Law has been a background assumption
which is under-theorized. Critical researchers assume that law is a peripheral force unable to create change. Uncritical researchers assume that legal
reform is an indicator of positive change. In this paper, I explore the relationship between legal reform and grassroots participation. In particular, I
examine the impact of two laws specifically designed to empower grassroots groups to participate in ecological conflicts. I argue that law is important to understand because it plays a large role in shaping the experiences
of grassroots groups as they seek to participate in environmental conflicts.
I contend that critiques of legal reform as being merely reformist miss the
way the law shapes environmental conflicts. Equally problematic are
uncritical assessments of legal reform as being indicators of progress.
These assessments miss the way that legal reform places new barriers in
front of disempowered groups. I conclude by suggesting different theoretical strategies that can be used to craft legal reforms that would support
grassroots organizing.

The Treadmill of Production as a Conceptual Apparatus

Any understanding of environmental conflicts has to start with some
notion of the relationship between environment and society. The concept
of a treadmill of production symbolizes the basic dynamics of natural
resource use in the United States and, increasingly, in other industrial and
industrializing societies (Schnaiberg 1980, 1986, 1994; Schnaiberg and
Gould 1994; Gould 1991, 1993; Gould, Schnaiberg, and Weinberg 1996;
Pellow 1994; Kousis 1991, 1993). The dynamic theory associated with this
concept can be summarized as follows: Ecosystems are a pivotal part of
market-based political economies because they can be converted by partic-

ipants into profits through market exchanges. These profits can then be reinvested through the purchase of productive physical capital into more profit. In turn, successive stages of profits can then be used to purchase more productive physical capital rather than investing in improving labor, enhancing environmental protection, or increasing social security. At each stage of this acceleration, such financial and technological change raises the capital-intensification of production, often lowering its employment capacity and locking the organization into a need for a constant and reliable flow of natural resources.

Such reductions in employment and social security potential per unit of production, in turn, provide social and political support for enhancing further acceleration of the treadmill from non-investing population segments. Thus, both producers and state organizations become increasingly committed to and dependent upon ensuring the growth-values needs of producers. Producers need assurance of reliable, consistent, and inexpensive access to natural resources in order to ensure the efficient and profitable use of capital. Simply stated, the capital intensification of the production process requires profitability to repay outlays and a reliable, accessible supply of natural resources to keep production growing to generate new capital. The state needs the same assurances to ensure that national production is generating enough surplus to: (1) support outlays to capital owners, (2) provide enough additional exchange values and social surplus to supply an adequate level of wages to maintain consumer demand, and (3) generate enough tax revenue to cover social expenditures of the state.

While the emergence of some ecologically sensitive production practices challenge some notions of the treadmill (Mol 1995), the overall dynamics of the marketplace are still captured with this concept. In particular, the theory highlights that within a market-based economy: (1) there is continual pressure to always accelerate the process (for example: higher profit margins, more efficiency, increased growth), and (2) as this occurs, there is a widespread social belief that we are locked into "this way of doing things." The only way to reduce the social risks of being dropped from the benefits of the treadmill appears to be to speed the treadmill up. Thus, the marketplace has the following goals: expanding industrial production and economic development, increasing consumption, and maintaining a political alliance of private capital, labor, and governments to promote these goals. When the treadmill is the dominant model, improvements in public welfare become tied inextricably to economic growth.

As participants become increasingly more dependent for their livelihoods upon economic expansion, ecological disruption occurs. Given the basic laws of thermodynamics, most forms of economic expansion entail both a depletion of natural resources (withdrawal—reflecting the Law of Entropy) and contamination of ecosystems (pollution—reflecting the Law of Conservation). The historical dynamics of the marketplace, juxtaposed against the basic scientific principles of thermodynamics, leads to a scarcity of natural resources for more humans and animal species. Environmental conflicts are fundamentally battles over these various forms of scarcity.

Four groups routinely join these battles. Producers seek to maintain ecosystem access. They are highly conscious of their material need for natural resources and use their assets to capture exchange values in markets, which are defined as the transformation of ecosystem elements through industrial processes, usually for market exchanges. They also try to influence the state to maintain the flow of, and access to, these resources. Environmentalists fight to slow down the treadmill, reform it, or dismantle it. They seek to maintain use values of ecosystems, which are defined as the relatively direct utilization of ecosystems by individuals, families, and communities without substantial alteration or transformation of the ecosystems. Communities have mixed needs and are highly conscious of use-value interests, including access to natural resources for either biological sustenance or recreational purposes. Contrarily, communities also have exchange-value orientations. As workers, their companies depend upon access to resources; as taxpayers, they worry about the costs of safe waste or sewage disposal and the costs of growing local regulatory bureaucracies. The state is both a facilitator of capital accumulation and a social legitimator of the socioeconomic structure for the citizenry (O'Connor 1973, 1988). In its role as facilitator of a prosperous economy, the state needs unlimited access to natural resources for exchange values. In its role as distributive justice legitimator, it must maintain resource levels for use value.

Environmental conflicts arise when these four groups struggle to control and shape the access of ecosystems within the treadmill of production to fit their mixture of use and exchange values. The history of these conflicts is one of emerging dialectical conflicts. The dialectic has two parts: (1) with few exceptions, ecological systems cannot meet both exchange-value needs and use-value needs, and (2) the treadmill of production places primary emphasis on exchange value. However, it is use values that are a social necessity for all classes and groups. Use values are shared by a much broader social and political constituency, but a less powerfully organized one than are exchange-value interests (Schnaiberg 1994).

I have argued in a recent book (Gould, Schnaiberg, and Weinberg 1996) that grassroots groups seeking to push for ecological concerns face three primary problems. First, participating in an environmental conflict requires tremendous resources. Grassroots groups are seeking to participate in highly charged political conflicts where groups have to be able to combat extremely powerful producers and state participants. To do this, they need to mobilize technical expertise (lawyers, ecologists, or toxicologists), financial resources, political capital, time, and local support. Second, grassroots groups often find it extremely difficult to articulate their concerns for ecological protection in a legitimate way. Conflicts are inevitably reduced to an issue of economic growth, and views for ecological protection are labeled naive, simplistic, irrational, or a luxury. And third, environmental conflicts tend to be multidimensional. While local grassroots groups may be trying to act locally, the locale is often multidimensional, stretching across local, regional, and global political arenas. Groups are thus required to open public space for themselves in places that are far removed from the local terrain. In combination, these factors often leave grassroots groups exhausted. In interviews, participants express amazement at the personal cost of grassroots organizing. The question addressed here is: Can legal reforms mediate these problems?

Research Design

The research to be described began as an analysis of 41 attempts by grassroots groups to participate in local environmental conflicts. These groups comprise the full population of two projects run by the Sierra Club. From 1989 to 1994, the Illinois chapter of the Sierra Club (hereafter referred to as the Sierra Club) implemented a grassroots strategy. Prior to 1989, the Sierra Club had pooled local resources through dues and phone banks to lobby for better regulations and to help elect environmentalists to various political offices. It was a strategy that sought to create a strong, centralized, formal organization that could exercise political and legal power. By the mid-1980s, the Sierra Club was forced to react to a political environment that was increasingly hostile to environmental protection (Grieder 1992; Schnaiberg and Gould 1994; Shabecoff 1993). Following the lead of other environmental organizations, the Sierra Club began to decentralize both its resources and its activities, creating numerous, small, informal organizations (Shabecoff 1993; Dunlap and Mertig 1992; Szasz 1994; Cable and Cable 1995). This grassroots strategy was summarized by a prominent Chicago-based environmental activist: "The problem isn't laws, it's enforcing them. We need to train ourselves to use our rights to hold industry accountable to the laws we got passed in the 70s....We're going to be the eyes and ears of the EPA....We've got to get people in every area (township) doing this stuff." The new strategy rested upon the premise that small, informal, locally based organizations could get active locally, while the combination of action in many locales would make a difference globally (Gould, Schnaiberg, and Weinberg 1996).

The first project, the Swamp Squad, was an attempt to preserve wetlands by training grassroots groups to use Section 404 of the Clean Water Act of 1972. Section 404 prohibits the discharge of dredged or fill material into wetlands without a permit from the Army Corps of Engineers. Section 404 also requires that all permits be given final approval by the Environmental Protection Agency (EPA). The Sierra Club ran training seminars outlining the 404 process. These seminars informed grassroots groups about the 404 provisions for public comments on permits. It also stressed secondary regulatory systems (for example, village and county zoning processes) that also had formal procedures for public participation. After the seminars, the Sierra Club provided activists as mentors to groups that became involved in local conflicts.

Early experiences with the Swamp Squad led the Sierra Club to develop the Adopt-A-Plant Project (AAP). The Sierra Club trained local groups to use the 1986 Federal Emergency Planning Community Right-To-Know Act (CRTK). Under this Act, local groups had the legal right to know what toxic chemicals a particular manufacturing firm was emitting into a local community. AAP groups targeted local manufacturing facilities and, using CRTK, they gathered information on the health hazards posed to the community by the facility. Working with the Sierra Club, the groups used this data and other political strategies to shape local public discourses about appropriate production practices and to induce companies to reduce toxic chemical emissions.

The data on each of the 41 groups was systematically collected by using traditional qualitative techniques. I was the primary node for communication between the Sierra Club and the local groups—relaying information between the groups, organizing training sessions, tracking the progress of groups for the Sierra Club. I gained this depth of field access through my role as the Sierra Club's coordinator for both projects. I attended every

Sierra Club organizational meeting, workshop, and planning session, and I attended every event for both the AAP and Swamp Squad projects. With two exceptions, I interviewed every group during the initial mobilization period and continued to informally interview them at every stage of the conflict. The other two cases are Swamp Squad cases that occurred shortly before I took over as project coordinator. In these cases, I depended upon interviews and other materials (newspaper reports, internal documents).

HOW DOES LAW SHAPE GRASSROOTS EFFORTS TO ORGANIZE AROUND ECOLOGICAL CONCERNS?

A Case Study: The Residents of Downers Township

In the fall of 1990, a workshop was held in Downers Township, a small community northwest of Chicago. The workshop was organized by Janet, the editor of a local newspaper and a newcomer to the area. Despite the absence of heavy industry in the area, Janet was concerned about a few facilities that seemed to have constant—albeit small—emissions. A few months prior to the workshop, Janet and her assistant, Diane, had tried to write an article on the air quality of Downers Township. They made over 20 telephone calls to government agencies trying to obtain information; however, the calls yielded nothing. Staff members at City Hall assured Janet that the health of the community was fine: "Just look at Chicago," one person commented. The comment infuriated her. Chicago was known for its unhealthy air. It was not an acceptable benchmark. Likewise, Janet knew that many highly toxic chemicals were not visible. Officials with state and federal agencies had been even less helpful. A pattern emerged—one phone call would lead to a series of transfers. Every time Diane and Janet located somebody, the staff person was inevitably busy. "Frustration and worry…and a high phone bill" were the products of the first week. Week two started off better. Diane located somebody at the local Health Department who told her about Community Right-To-Know. This person was unsure of the details about the law but knew that companies were supposed to report their toxic chemical emissions. At Diane's request, he grudgingly agreed to provide total emission releases for each company. Unfortunately, he could not locate a breakdown by chemical or any corresponding health information. With this limited information, Janet and Diane set out to find some additional help.

With no idea of whom to call, Janet and Diane began by contacting environmental groups, all of whom appeared to be unfamiliar with CRTK or were not working on toxic chemical issues. Finally, Janet called the Sierra Club. The staff person at the Sierra Club informed her about the Adopt-A-Plant Program and provided the name of a contact person. Shortly thereafter, Janet called the project coordinator, and together they devised the idea of a workshop.

Six weeks later, a workshop was held, which 30 people attended. Janet introduced the two AAP volunteers (the project coordinator and another, the Sierra Club volunteer who had helped organize the project). The AAP volunteers proceeded to summarize the problem of toxic chemical pollution and the promise of CRTK. In Illinois, they explained, CRTK data could be collected from the Illinois Environmental Protection Agency (IEPA), the Local Emergency Planning Committees (LEPC), the Metropolitan Water Reclamation District (MWRD), and an EPA database. They explained what could be obtained from each of the sources and described the process of gathering the data. Fortunately, CRTK data were available for the companies in Downers Township, and this included a breakdown by chemical for each company. The project coordinator immediately displayed a folder and began to recite some of the more interesting emission data. Unfortunately, the scope of the data was restricted. The project coordinator explained that CRTK data captured only a small fraction of the toxic chemical emissions in any locale, and only approximately 330 out of the estimated 60,000 chemicals used by private industry fell purview to the law. Furthermore, various loopholes existed that exempted many companies; for example, calculation techniques allowed companies to hide emissions. Finally, there was no enforcement. Combined, these shortcomings were significant. For example, the U.S. General Accounting Office had reported that in 1987, up to 95 percent of all chemical emissions were exempted from reporting. Some facilities had been able to hide an estimated 89 percent of their emissions in 1987 while still complying with the law. Additionally, some companies were using false emission data to stifle local mobilization efforts. The project coordinator suggested that Diane use the CRTK data to stir up local interest. Janet found this idea intriguing: "What if we follow up the article with a workshop? I could announce it at the end of the article." The project coordinator agreed that this was a good idea.

An hour into the meeting, Janet suggested that they take a five-minute break. Immediately, a group surrounded the AAP coordinators. The group included: Gary, a high school teacher; Susan, a lawyer; Susan's husband, Ed, an ex-EPA employee; Frank, a retiree and long-term resident of Downer's Township; and Sally, another local resident. Gary stated that a few miles from the library (where the workshop was taking place), a company (which I will call International Producer) was located. For years, residents of Downers Township had been concerned about the dark smoke, foul odors, and dirty streams that seemed to emanate from International Producer. Periodically, various groups of people had tried to place pressure on the city to "do something." Unfortunately, their efforts never amounted to much, because residents had never been able to obtain any information about International Producer. The AAP coordinator responded that usually sight and smell were revealing. Susan laughed and said, "But they (International Producer) can make us look stupid and uninformed." Gary responded that actually International Producer had spent years ignoring the community by arguing that "we're following the law." Sally commented, "They're always in compliance, but we can't see across the street [because of smoke]." Gary explained that the city was "in their back pocket, and also the guys who work there worry about their jobs." The other project coordinator (who worked at one of the federal agencies and had toxicological knowledge) suggested that they examine the CRTK data that the project coordinator had brought. She glanced at the data for International Producer and announced to the group that two of the chemicals listed as being emitted in large quantities had particularly dangerous (carcinogenic) health effects, thus assuring the workshop attendees that International Producer had toxic emissions that were worth worrying about.

When the workshop resumed, the project coordinators asked if people were interested in creating an AAP group. Eleven people responded affirmatively. The project coordinators proceeded to describe efforts by other AAP groups, including developing public relation campaigns, and suggested programs that local polluters could develop to reduce the use and emissions of toxic chemicals. Mostly, it was explained, AAP groups were using the list of known chemicals (complied from the list of CRTK emissions) to locate "safer and less expensive" alternative chemicals. This sometimes included learning what other companies in the same industry were doing to use fewer dangerous chemicals. Groups were also using CRTK data to publicize the problem of toxic chemical pollution, thereby attracting support from the community. The goal of these efforts was to get companies to sign Good Neighbor Agreements which were nonbinding commitments to respect the rights of the community. Janet spoke first. She had heard rumors that the township was negotiating with the company over the height of a new emissions stack. Ed suggested that this would be a good time to attempt to open a dialogue with International Producer, as the company would be sensitive to community support. The group agreed to spend the next month gathering information on the company.

A month later, the group held their second meeting. This time the AAP coordinators brought with them the CRTK person from another regional environmental organization (REO) that often also worked on the AAP. Immediately, it was apparent that everybody had run into great difficulty in their attempts to gather information. The CRTK figures that the Sierra Club had provided were from 1987. Janet and Diane had been given the task of obtaining updated emissions information on International Producer. They found this task difficult. Even though International Producer had complied and reported 1988 data, it would take the EPA another four to six months to make it available. The data from 1989 would not be available for another year. Beyond being old, the 1987 data were too sketchy to be of much use without the permit and compliance records. Susan, who had attempted to locate the permit data, reported that this was impossible. Permits were held by too many different regulatory agencies. First, there were federal, state, and local permits, each held by appropriate agencies at each level of government. Second, at each level, different permits were held by different agencies. She had received some information from the Illinois Environmental Protection Agency (IEPA), but it was "impossible to read (the photocopying was dark and wavy) and then I couldn't understand what the form meant." The REO representative examined the sheet and informed her that it was CRTK data that confirmed the data they already had. When Susan laughed, the AAP coordinator responded: "Actually, we needed this. The figures I gave you were from a computer database that has lots of typos on it." Next, Gary, who had been assigned the task of obtaining information from the LEPC, described how he played "telephone tag" for three weeks and was unable to establish a mutually convenient time to contact the LEPC person, who was a volunteer. Everybody expressed frustration. "It's impossible to wait by the phone all day." "I can't go traveling to Springfield to get this stuff [information]." "If we do get it is it worth it?"

Having listened to the reports, the project coordinators suggested an alternative plan. It was apparent that the process of gathering data was depleting the group of scarce resources. Each participant had remarked that they were dejected. "We have enough information to contact them. I mean we know enough to make a convincing case that there is a problem of toxic pollution." However, without the CRTK data, all the group could do was make angry statements about smells and black smoke. With the CRTK data that they had gathered, they could make statements about emissions of toxic chemicals that had known human health effects. This gave the group enough information to approach the company about an admitted problem (since the company had reported the figures) and to make statements that were somewhat educated. Given this, the group decided that they would meet with the company.

The next day, Gary contacted the plant manager at International Producer, who claimed that the company was in compliance with every law, that the citizens would try to steal trade secrets, and that he was too busy to spend a day with the group. Gary spent an hour on the phone convincing the plant manager that the group did have something to offer. Working through the Sierra Club and REO, the group could offer technical support to the company on cost-effective emission reduction schemes that would save the company money, while creating goodwill between International Producer and the residents of Downers Township. Furthermore, the group would not steal trade secrets. Nobody in the group worked in the same industry. Finally, the group acknowledged that International Producer was in compliance with its permits. The group was not accusing International Producer of anything illegal. They only wanted to establish lines of communication between the community and the company. The plant manager responded that he was skeptical. Furthermore, he was not under any obligation to appease a group that had neither a right to trespass nor to stand in the company's way of staying profitable. The phone conversation ended with the plant manager agreeing to "give it some thought."

After consulting with the project coordinator, Janet sent the manager a follow-up letter (that was adopted from the AAP manual), which stated in part: "From the onset, I want to stress that our interest is in opening a friendly discourse. We respect your right as a valuable member of our community." After receiving the letter, the plant manager agreed to meet with the community group under the condition that only six people from the group would attend and that only two could be the Sierra Club staff. The company would set the agenda and the group would have to adhere to it.

Six weeks later (it was hard to find a common time), the group met with the company. From the start of the meeting, the atmosphere was tense. The group was ushered into a formal meeting room. The company representatives went through a three-hour prearranged monologue covering what the company did, what their pollution prevention program consisted of, and how their latest ("just reported") CRTK emissions levels had decreased from the figures that the group had obtained from the EPA. At various times, members of the AAP group tried to ask questions, but each time the plant manager remarked that there would be times for questions afterward. After three hours, the plant manager, as one citizen put it, "then informed us that the meeting was over, and got up and left." The group followed him out into the hall inquiring about time to ask questions. The plant manager responded, "I am sorry but this went longer then we anticipated." The group asked if they could meet with the company's "waste minimization team" and offer cost effective strategies. The plant manager replied, "You have no technical expertise." When the group responded that they did in fact have technical experts, he again accused them of trying to steal trade secrets. When they asked if they could at least meet with him again, he responded: "What would the point be?" As a last resort, the group handed the plant manager a detailed, 10-page questionnaire with 35 questions and asked if he could look it over and meet with them. Again, quickly glancing over it, he claimed that everything had been answered and abruptly handed it back to them.

Over the next few weeks, the group tried to contact the plant manager, but he refused to answer phone calls or letters. It became clear to the group that they had to change tactics and become more aggressive. Friends who worked inside the plant informed group members that the company's plan of action was to stall, hoping that the group would tire and disappear. The company had originally met with the group because they were curious about what the Sierra Club and REO were "up to." According to these friends, the company felt strongly that they had no obligation to "waste time with you [the group]." International Producer was in compliance with every permit. CRTK required them to report figures, a process they had not only complied with but one they had superseded by meeting with the group. Nobody at the plant felt that the group needed to be appeased any more.

The project coordinator called various public officials to see if they would lend support. He learned that the company was negotiating directly with regulatory agencies and the city, and neither of those groups were going to risk alienating International Producer by working with the AAP group. When pushed, contacts responded that they saw no reason why the company should appease the group any more than it already had, and the group had no right to take additional steps, especially since International Producer had already gone farther than the law required. It became apparent that International Producer had no intention of meeting with the group again, answering follow-up questions, or letting the group conduct a plant visit. Furthermore, nobody was interested or saw a reason why the company should.

The group never met again.

How Procedural Law Shapes Environmental Conflicts

While the conflict between the residents of Downers Township and International Producer was political, the law shaped it. As represented by CRTK, American law is distinctly liberal. It arises from a belief in the classical liberal doctrine of free market capitalism, minimal government, and a jurisprudence of strict interpretation of fundamental law (Anderson 1990). Embedded in this conception of law are notions of process. The doctrine of liberal law is to provide a framework that allows individuals to freely choose and pursue their own needs and desires (Anderson 1990; Kennedy 1976). The project of liberal law is to create and maintain the stable, comprehensive frameworks that individuals need to develop these plans and projects. Process is paramount because individuals can do this only as long as decisions and rules emerge in a consistent way that allows them to draw on past experiences to gauge future options and to predetermine the likely outcomes of current practices. In practice, this necessitates that the law rest its decisions on impartial and general principles, consistently applied across contexts. Another way to think about this is that law is conceptualized as concerned with procedures and process to the detriment of substance and outcomes. Principles are reconciled through processes, making the domain of law a set of rules not standards (Friedman 1975). No one person has institutionalized privilege over another, but rather the invisible hand moves society as individual interest is reconciled into a common good.

There are some disparate strands here. First, the process ensures the consistency of rules and standards needed for individuals to plan. Second, the process ensures that information flows in ways that ensure that individuals can gauge interests. Third, the process ensures that competing interests can engage in market-like behavior such that social equilibrium points are achieved.

CRTK was a procedural law. The concept of giving a community the right to know about toxic chemical pollution could have meant a variety of things. For example, the law could have been outcome-oriented, with the EPA taking clear steps to ensure that every community member knew and understood what types of contaminants existed within their community. The EPA did not do this. Instead, it started with the unstated assumption that CRTK was not to be designed to achieve an outcome, but rather that CRTK would provide a framework within which individuals could freely chose and pursue their own needs and desires (Hadden 1989). The EPA defined "the problem" as an issue of information. Having information would allow communities to make rational decisions about present and future actions. The EPA defined its role as mediating this process. The EPA would serve as an objective scientific body that could locate, store, and disseminate the data (see, generally, Landy, Roberts, and Thomas 1990; Yeager 1991; Hawkins 1984). To do this, CRTK becomes a law of process. CRTK mandated that various governmental agencies enact a variety of processes. In our case, Downers Township was charged with constructing a Local Emergency Planning Committee (LEPC) comprised of representatives of local government, emergency and disaster services, local industry, local community groups, and the press. The LEPC was mandated to store data on the type, amount, and location of toxic chemicals at local businesses. Included in this would be Material Safety Data Sheets (MSDS), which documented the name of every hazardous chemical used at the plant and the hazardous component(s) of each chemical, and Chemical Inventory Forms (Tier I and II forms), which contained an estimate, in ranges, of the health hazards and physical hazards of the maximum amount of hazardous chemicals on site, the average daily amount present on site, and the general location. Furthermore, International Producer and other local businesses were required to notify the LEPC of all accidental toxic chemical releases that occurred and to follow these reports with an emergency notice that described the nature and health risks associated with the release. The Illinois Environmental Protection Agency (IEPA) would buttress the data collection by storing Toxic Chemical Release Forms (Form R's) which documented what toxic chemicals were present, how they were used and disposed of, and an estimate of yearly total amounts present at the facility. All of this information was to be available to groups like the residents.

CRTK, molded as a piece of liberal legislation, was concerned with procedures and process to the detriment of substance and outcomes. CRTK did not attempt to codify a set of standards but rather to create rules for giving information. CRTK did not require that the EPA ensure that a percentage of the public actually become aware of the data. The law contained a set of processes that made it possible for this to happen but, ultimately, principles were reconciled through procedures, making CRTK's domain rules not standards (Friedman 1975). Never was there any question or concern about whether the residents could actually use the information provided, much less if they could actually access the information. The law was not concerned with direct outcomes or particular cases. Nobody cared when the residents complained that they had been unable to reach the LEPC contact or that nobody knew where to obtain CRTK data. The response of a state official was: "The process is in place, we cannot be concerned with public outreach." Nobody followed up to make sure that the information obtained by the residents was clear or that they were able to use it. Nobody cared because these outcomes fell outside the purview of the law.

By being procedural, CRTK shaped the conflict in three basic ways. First, the process of mobilization was transformed. The group had mobilized by learning about and accessing legal process. Janet and Diane had to learn that the law existed. They had to learn about the processes of acquiring CRTK data. Also, they had to interpret the CRTK data. Much of what the group did was shaped by these tasks. For example, most of the work had to be done during the day when the IEPA was open and technical experts could be found. This determined who was able to join the mobilization effort, what those who mobilized did, and when they were able to do it. Second, other peoples' reactions to the group were shaped by law. In particular, other people defined their reactions to the group by "what the procedures required." The procedural requirements of CRTK mandated only that objective information be presented. Nobody at the IEPA had to do anything more than photocopy a sheet of paper that contained some basic information about levels of toxic chemical pollution. When pushed for more information, IEPA staff used the mandate of law to adopt a "removed objective" persona. Third, law shaped the types of arguments that the residents of Downers Township felt compelled to make. Across each of the arenas in which this conflict took place (the media, the EPA, the meeting), the arguments that each actor made were about having the legal standing to take an action. The residents argued that they had a legal right to safe air, while International Producer argued that they had a legal right to engage in commerce. The EPA and local zoning boards, although more peripheral in this case, made the argument that they had the legal right and duty to issue permits. No actor made a political argument about strength or a social argument about equality. Thus, International Producer did not argue that they would continue polluting because they were "bigger and stronger." Likewise, the residents did not argue that reducing emissions or granting access to the facility was the fair thing to do. Each of these actors may have believed these statements, and expressed them in private, but the public arguments made were about why a particular actor had the legal right to take an action.

DO GRASSROOTS GROUPS SUCCEED BECAUSE OF LAW, OR DO SUCCESSFUL GRASSROOTS GROUPS OVERCOME LAW?

Obviously, laws are important; basic rights are essential to any democratic political process. To paraphrase Patricia Williams, the only people who do not appreciate basic rights are members of dominant groups that do not need them (Williams 1990). Having stated the obvious, it is important to examine how procedural laws tend to impact grassroots groups. One way to state this question is as follows: Do grassroots groups succeed because of laws or do successful grassroots groups overcome laws? The answer is mixed. Laws provide grassroots groups with beneficial pieces of information, and also provide groups with clear processes so that they can participate in decision making processes, but successful groups often do not have the means to use the information or recourse. Thus, they succeed without really using the laws and, furthermore, procedural laws can exacerbate pre-existing problems.

In general, Swamp Squaders and Adopt-A-Planters groups succeeded in one of two ways. Many groups succeeded by creating alliances with powerful decision makers. These grassroots groups often lacked rudimentary knowledge about the issues and conflict. They had no knowledge about the proper ways of accessing legal proceedings. They did not know who the other participants in the conflict were. These groups did, however, have resources. In particular, they had a strong leader who was able to mobilize community support and technical expertise. These groups succeeded when they were able to use this local support and technical expertise to create alliances with decision makers. Adopt-A-Plant groups created a relationship with a plant manager, while Swamp Squaders found allies on zoning boards or at the federal agencies (Army Corps of Engineers or EPA). These groups basically traded their political support to an ally in return for a chance to become part of a decision making process.

Law was helpful to the extent that it gave them either a piece of information or a place to use information. AAP groups were able to acquire information about local toxic chemical pollution. This information provided them with a list of local producers with particularly bad track records. This list was also helpful in gauging which companies were most important to mobilize against and helped the AAP identify what companies might be most concerned about negative media coverage and, thus, might be most willing to trade political support for access to internal decision making processes. Swamp Squaders were able to locate a hearing or process where they had a legal right to participate.

While CRTK and Section 404 were useful in these ways, they were often less useful than might be apparent. First, the data provided was often information that groups already had in the form of local knowledge (Geertz 1983). AAP groups already knew which local facilities were emitting toxic chemicals because they saw it coming out of smokestacks and smelled it in their homes. Similarly, Swamp Squaders usually learned about a proposed development and then had to search for the permit application. The process rarely worked the other way around. This is a crucial distinction. One of the touted benefits of the Section 404 procedure was that developers had to apply for permits to fill in a wetland, thereby making the information available to local community members. In practice, the process rarely worked this way. Swamp Squaders almost never learned about a proposed development project from a permit application. Permit applications were publicized in information channels far removed from community groups.

Second, these laws were somewhat beneficial, but often not really usable. For example, CRTK gave grassroots groups more precise information about pollution, but groups rarely used it. Most often, they could not decipher the data. When they could interpret the information, they found that it never became relevant in any public discourse. CRTK gave them technical detail about pounds of pollution; however, public discourse was rarely about pounds of pollution. It was either about health risks associated with exposure or about the relationship between pollution and various forms of economic growth (for example, potential loss of jobs from more regulation or decrease in property values associated with pollution). CRTK provided little information that was relevant for these types of debates. Section 404 often gave groups permission to write a technical objection to a permit that they had neither the expertise nor the information to write. Once again, the information provided by the law was not the information they needed in the larger political conflict.

Third, CRTK and Section 404 exacerbated preexisting problems. The largest problem faced by Swamp Squad and Adopt-A-Plant groups was trying to track a conflict through a maze of political and legal processes. Swamp Squaders would have great difficulty tracking conflicts from one legal venue to another. A given conflict could go from multiple zoning boards, the Army Corps of Engineers, the EPA, a county board, a couple of township boards, or state agencies like the Department of Transportation and the Department of Conservation. Adopt-A-Plant groups would spend days trying to locate which people in what agencies had the specific information they needed. This meant that groups would only be able to use the law if they had access to preexisting networks that allowed them to track the conflict.

Other groups succeeded because they were able to reframe a conflict. Often, these groups already had both resources and information, which they used to identify an ally early in the process and to convince the ally that being environmentally responsible would help the organization meet its goals. In some respects, access to the legal process gave other participants a reason to listen to Swamp Squaders and Adopt-A-Planters. Groups made arguments about having legal standing, which created uncertainty for organizational actors. Producers and developers would listen to a group in order to reduce the uncertainty of political mobilization. Groups used this opportunity to make arguments about economic profitability through environmental quality. In doing so, they altered the construction of the conflict in a way that allowed them to participate. Adopt-A-Plant groups, for example, found that their legal right to obtain CRTK information created tremendous uncertainty for plant managers. Plant managers were worried that community groups would use the information "incorrectly"—either misinterpreting it or becoming needlessly "hysteric." To reduce this uncertainty, plant managers listened to and worked with the groups. This gave the groups a short but significant time frame to convince the plant manager that (to quote a letter from one group), "[w]e want to work with you to ensure profitability through environmental quality." A similar process occurred with Swamp Squad groups. Their right to comment on permits created uncertainty for city officials and developers at a late point in the development process (after the city had invested significant time and financial resources into the project). Again, this created a brief moment for local groups to create ties if they argued that they wanted to ensure economic health through environmental quality.

It is important to recognize that law also mediated their voices. Section 404 and CRTK severely restricted what Swamp Squaders and Adopt-A-Planters could argue. Section 404 mandated that groups argue about technical feasibility of a project or ecological quality of a contested wetland. This placed tremendous burdens on a group to gather the scientific expertise needed to make such a technical argument, and it also meant that groups could not legitimately argue about quality of life concerns. Much of what they considered environmental was deemed non-pertinent to the legal questions. Adopt-A-Plant groups had to convince producers that complying with the CRTK law was not enough. Plant managers often argued that they did not need to meet with groups or worry about pollution because they had already complied with "the law."

Obviously, grassroots groups are better off with CRTK and Section 404. There is nothing in these cases to suggest that the efforts would have been more successful without the laws. However, it is also important to remember that most of these groups would have succeeded without the laws in place. CRTK and Section 404 made it easier, but they also made it harder. Most importantly, the laws were often tangential to the conflict because groups were not able to use them.

Take, for example, the following case. An Adopt-A-Plant group emerged in an area where local residents were concerned about toxic chemical air pollution emanating from local manufacturing plants. The Sierra Club located rudimentary information indicating that several of these companies were extremely sensitive about protecting their local image. In fact, most of the companies had been working through their trade associations on a "Responsible Care" program. The group also learned, again through the Sierra Club, that a number of these firms were "facing tough times from global competitors." The managers could not afford local mobilization, which could increase costs and reduce consumer support. Finally, the group had preexisting ties with people in the facility and in the local political community. Initially, the plant manager was reluctant to meet with the group. He argued that the company was complying with the law and did not have to meet with the community group. The AAP volunteer responded by arguing that the group had information that would save the company money. In a follow-up letter, the group stated, "We want to work with you to find win-win situations." Finally, the plant manager reluctantly agreed to meet with the group.

Working through the Sierra Club, the Adopt-A-Plant group located a technical expert who accompanied them on their tour of the facility and subsequent meeting with company officials. In addition, the group devised a list of steps taken by similar facilities in other areas to reduce the amount of toxic chemicals used, thereby saving money and reducing emissions. With this technical help and the CRTK data, the Adopt-A-Plant group approached the plant manager from a position of strength. They claimed a legal right to obtain information, political strength to use that information ("we represent the local community"), and knowledge ("we have technical expertise that can save you money and political trouble"). They convinced the plant manager that it would be beneficial to all parties to work together to "develop a cost effective toxic chemical use reduction program." Prior to the meeting, the plant manager convinced others to listen to and respect the group. The Adopt-A-Plant group used this opening to reframe the issue. Instead of constructing toxic chemical pollution as an issue of "how we can hide emissions to maintain exchange values," they reconstructed it as "lowering our emissions is a sign of efficient chemical use." At a follow-up meeting, the plant manager stated, "When we realized the cost of our emissions, we were staggered, those emissions are lost profits....Sometimes we want to do the right thing, but need you all to show us how to do it." The company agreed to take many of the steps the group suggested and even went farther in a number of instances. They invited the technical expert and a member of the group to sit in on internal environmental committees. The Adopt-A-Plant group had been able to reconstruct the issue as "this is one of those unique situations where we can have economic growth through environmental quality."

In this case, the group succeeded but only tangentially did it use CRTK.

BLINDSIDED BY THE LAW: THE UNSUCCESSFUL CASES

Most of the time grassroots groups do not succeed in participating. Of the 41 groups in the present study, 25 were unable to participate in a chosen conflict. Groups failed because they faced tremendous power inequalities vis-à-vis other participants in the conflict. This is not a surprising finding. It is consistent with the environmental literature and, more generally, with the social movement literature (see, e.g., Piven and Cloward 1979; McAdams 1982). A more important observation is that law plays a role in exacerbating these power inequalities for unsuccessful groups: It is partially to blame.

Groups often failed because they lacked basic information about a conflict or issue. Often, these groups had surprisingly high levels of resources. Swamp Squad groups frequently had strong local support for saving a wetland but never had good information as to when a wetland was "threatened." They were always one step behind a conflict. For example, two groups learned about development projects *after* the wetland had been destroyed. In both cases, the groups had actively been engaged in trying to save the wetland but did not learn about a proposed project until the wetland was filled. Adopt-A-Plant groups had similar frustrations. Most often they were denied requests for meetings and plant tours, or Freedom of Information Act letters were accidentally "lost." They were basically ignored.

In many instances, CRTK and Section 404 were used against these groups. Using CRTK and Section 404, EPA and Army Corps staff drew clear boundaries around "what I am required to do (by law)." Officials often denied local groups audiences by claiming, "They can participate in the 404 process just like anybody else," or "Why should we give them special treatment?" Developers and plant managers adopted a similar stance. One plant manager refused to acknowledge the Adopt-A-Plant group, clearly stating, "Right-To-Know (CRTK) says I have to give the EPA information, there's nothing about meeting with you." Laws gave unresponsive bureaucrats and producers a way to set clear boundaries between what they *had to do* and what they *did not have to do*.

For example, a member of one local neighborhood association read in the paper that the Park District had decided to purchase a wetland behind her home. The Park District had been considering the acquisition for a long time. Members of the association had recently stopped attending the Park District meetings, because the frequency and length of the meetings made it impossible for them to commit to attending on a regular basis. Shortly thereafter, the Park District announced that the wetland would be turned into a recreation park (complete with paddle boats, a running track, soccer fields). Working with the Sierra Club, the Swamp Squader mobilized members of the association, and together they devised a detailed proposal outlining the reasons why this purchase was a poor use of Park District funds. Over the next week, the neighborhood association attempted to locate people within the Park District Board who might favor their alternative strategy to preserve the wetland. When they approached the Park District, they were informed that the issue had been resolved and was waiting city council approval, and the meeting would take place the following night. Immediately, the Swamp Squaders called the Sierra Club. Working with the Swamp Squader coordinator, they again developed a strategy and outlined an alternative to present at the meeting. Still, they knew that the board viewed wetlands as wasted property that brought in neither tax revenues (like homes and businesses) nor votes (like new parks and recreation areas). The meeting was a disaster for the group. As one person commented, "We spoke but nobody even heard....I don't think that one person even looked at us." When questioned by a Swamp Squader after the meeting, an angry board member commented, "We followed the procedures. You had your chance to voice an opinion. These things are governed by very strict guidelines that we have to follow in making a determination." Another member commented, "My colleagues don't want to listen. Call it professional arrogance, but they don't think you guys have anything to offer....They will use any procedural rule to claim that you had a chance."

Similarly, plant managers used CRTK to stifle mobilization by pulling out data that "were not yet reported" or "were projected for down the line." This newer information made the concerns of Adopt-A-Plant groups appear to be alarmist or naive because it always showed declining emissions. In all of these cases, groups exhausted their resources trying to work their way through legal processes. This was particularly troublesome for Swamp Squaders who found that Section 404 required them to comment on permit applications in a technical manner which necessitated time, technical training, connections, and financial resources.

Other groups were unsuccessful because they were never able to mobilize enough resources to use good information that they had acquired. For example, a Swamp Squad group learned about a wetland that was to be illegally destroyed the next day, but they were unable to obtain a response from the Army Corps of Engineers or EPA. An EPA official commented, "Can you verify the high grade quality of the wetland?" This was clearly a technical requirement that few Swamp Squaders had the skills to carry out and that would take weeks to perform. More commonly, an EPA official would comment, "The wetland is too small to visit or care about." In some cases, a phone call was not returned until the wetland was filled in.

Adopt-A-Plant groups could often obtain good data about toxic emissions but could never get anybody to listen to them. For example, two members of a group agreed to coordinate an effort against a local chemical plant. The facility was selected primarily because the CRTK data revealed that it had high emissions levels of two extremely toxic chemicals; moreover, the facility was located close to a new suburb and a school. Fortunately, one of the group members knew a mid-level manager at the facility, and a few weeks into the project, he had a long conversation with the insider. The conversation was discouraging. The parent company of the chemical plant was placing pressure on the facility to increase profits. The facility had internalized this as "cut costs wherever, whenever, however." Requests to cut pollution would be "laughed at by management." The company line was, "if we can show greater profits, then maybe we can indulge the Sierra Club's concerns." Working with the Sierra Club, the group tried to set up a meeting anyway. They sent a letter to the plant manager explaining that they had suggestions as to how to cut costs using fewer chemicals. A week later, they followed up with a series of phone calls that were not returned. When they finally reached the manager, he explained that the company was not interested; they were too busy. When the Adopt-A-Plant leader pressed harder, the manager replied, "Look, we're complying fully with every law." The Adopt-A-Plant leader responded: "We are not concerned with the law, we want to see if we can find a win-win situation." The plant manager responded that the company would not meet with the group, "We already devote too many resources to regulatory compliance."

In the end, some groups failed because the political construction of the issue negated any role for environmental protection. In each case, the issue was framed around needing to produce growth. For the Adopt-A-Plant groups, it was a matter of unlimited production. For the Swamp Squaders, it quickly became an issue of development. In either case, natural resources were thought of as commodities needed for the producers' growth and for local tax base stability. Producers and elected officials were united around the belief that "economic growth will allow us to purchase environmental protection." This framing placed substantial obstacles in front of groups; any arguments and forms of expertise that were not based directly around growth were viewed as distractions. In these cases, law channeled the discussion. Swamp Squaders were forced to argue cases in front of zoning boards whose primary concern was growth. Adopt-A-Planters were forced to argue with plant managers who were under tremendous pressure to increase profits. In all cases, others used the law to justify their actions. Producers and developers used the law to state that they had a right to engage in commerce, and nobody had a right to tell them what to do with their property or to prevent them from making money.

Across these unsuccessful cases, groups were blindsided by Section 404 and CRTK. I am not arguing that these groups would have succeeded if the laws had not been in place; this is clearly not true. However, it is also incorrect to state that these groups failed because the laws failed. In some important respects these laws were used, but to the detriment of the intended user. Other people used Section 404 and CRTK against the groups, labeling their concerns as outside the scope of the law or as requiring them to use certain legal process which ipso facto required them to mobilize resources that they were not capable of obtaining.

CONCLUSIONS: CRAFTING LEGAL REFORM THAT SUPPORTS GRASSROOTS ORGANIZING

At two follow-up meetings to the Swamp Squad and Adopt-A-Plant projects, participants were remarkably hostile in tone when reflecting on

CRTK and Section 404. When I posed the question of whether they would had been better off without the laws, successful groups begrudgingly said "no," while the unsuccessful groups oftentimes replied "yes."

When talking about liberal law, it is too easy to be dismissive. We can either too critically assume that laws are merely reformist and write them off as legitimators of unjust systems, or we can uncritically assume that laws create positive change, even when they create barriers for disempowered groups to overcome. Ironically, the shift by major environmental organizations to a more localized mobilization strategy was partially in recognition that at some point more laws were being passed without much ecological impact.

My argument can be summarized as follows: Grassroots groups often have to overcome a lack of resources, the difficulty of articulating their concerns for ecological protection in a legitimate way, and problems of multidimensional conflicts (Gould, Schnaiberg, and Weinberg 1996). Liberal law often exacerbates these problems. Liberal law is operationalized as procedural law. CRTK and Section 404 are good examples; each gave grassroots groups the ability to participate in a process. Section 404 was a form of due process where grassroots groups were given an opportunity to participate in decision making. CRTK was a right-to-know law. Each of these procedural laws exacerbated preexisting problems because they channeled conflicts to very specific places. Grassroots groups often had difficulty locating these places. This was most true for the Swamp Squaders, who found conflicts to be channeled to very specific branches of the Army Corps of Engineers and Environmental Protection Agency, with the result that the Swamp Squaders often never heard about a conflict until decisions had already been finalized. If they could locate the site of a conflict, both Swamp Squaders and Adopt-A-Planters either found that accessing these processes required resources that they did not possess or drained them of scarce resources.

Finally, the legal process exacerbated the skewed relations between participants. Most social and political resources are distributed through the market. The net result is that producers, developers, and state actors have greater access to resources than do small, informal citizen organizations. Producers and state actors existed in organizations with stocked resources and preexisting ties, as by-products of previous conflicts or interactions, and were placed in the midst of regular information flows. For Swamp Squaders and Adopt-A-Planters, resources were nonexistent or scarce. In a procedural process that continually requires the mobilization and use of resources, producers and developers could outlast grassroots groups.

The question is, how can we progress beyond these dilemmas? Law is important. In its ideal, law should empower groups. I want to close by arguing that there are two options.

Option #1 would be to take procedural law seriously by moving towards an outcome-based procedural system. Too often, researchers have uncritically accepted the assumption that liberal law does not have an outcome associated with it. Conventional wisdom states that liberal law sacrifices outcomes for process—but this is not true. Liberal law does have a clear outcome embedded within it. The philosophical principle of liberal law is that divergent interests can be reconciled through objective principles into equilibrium outcomes. This principle is designed to ensure the consistency of outcomes needed for a free market economy and the competitive arenas of political participation needed for democratic forms of governance. Procedural law is the means to meet that end. Yet, too often we accept the means as the end. Liberal law will work only when we judge it by how well these outcomes are reached, and not by having process in place. In a sense, we already meet half of these outcomes. Legal processes do a remarkable job of ensuring economic actors that decisions will be based on rational principles consistently applied across cases. We have volumes that record these principles in forms of regulations. We also take great pains to write legal decisions that justify these decisions based on rules. We never measure the level of participation.

This might sound too abstract, but it is remarkably concrete. We know what keeps grassroots groups from participating. As such, we can theorize about the conditions that would allow them to overcome these problems. I want to postulate that, given the cases presented in this paper, legal processes are likely to meet the outcome of participation when the following three conditions are met: (1) information flows throughout the community, (2) participants have some basic level of resources to competitively engage each other, and (3) the criteria for decision making legitimate a variety of views. Let me briefly explain each of these.

1. For a process to be competitive, people with an interest and stake in the outcome have to know that the decision is being made. Too often, community members have no idea that an issue has emerged or that there is a process of decision making that they can participate in. In cases of wetlands, the community generally has no idea that a wetland is being threatened until it is already filled in. In cases of toxic chemical pollution, the community often does not know that they

can actually do something about it. A working procedural law would have to be proactive in ensuring that information is flowing throughout the community, and state agents would have to have clear and active processes by which they inform community members about issues and procedures.

2. For competition to occur, participants have to be provided with the resources needed to compete with each other. We know what resources it takes: time, technical assistance, and access to media (Gould, Schnaiberg, and Weinberg 1996). The state can facilitate a more equitable distribution of each. Meetings can be held at night when community residents are more available to attend. Layers of regulatory processes can be merged so that single meetings encompass a variety of decisions. Technical assistance can be provided to interested participants. Just as we have legal aid, we need a model of legal/scientific aid where groups would have free access to qualified technical experts who could help them gauge impacts and construct arguments. Finally, the state can actively use its contacts within the media to publicize the concerns raised by community members.

3. Finally, for a process to adequately assess competing interests, it has to be open to a range of them. Oftentimes, ecological concerns are deemed illegitimate within a conflict. A process cannot promote competitiveness if half the competition is deemed illegitimate. Sometimes, the law does this overtly. In the case of Section 404, the EPA and Army Corps of Engineers are restricted in what they can consider. In the case of CRTK, the law refuses to take a stand on what should be done with the information. We can do better. We can essentially craft criteria that acknowledges the importance of ecological concerns.

Taking these suggestions seriously would mean different reforms for CRTK and Section 404. CRTK provides information, but few people know that it exists, fewer people can actually understand the information, and there is no formal arena for grassroots groups to use the information. Section 404 provides an arena but fails to inform the community about the decision-making process and denies members of the community access to the technical assistance needed to participate. Finally, both laws delegitimate the ecological concerns of grassroots groups. Overcoming these problems is not difficult. I have laid out three simple conditions that would have to be met. I am more skeptical that we possess the political will needed to meet these conditions. For a culture that talks a lot about the competitive marketplace locating equilibrium, we are remarkably hesitant to ensure that our competitive processes are actually competitive in any meaningful way.

Option #2 would be to move more closer to an outcome-based system of law. Here, the emphasis would be on crafting laws that establish outcomes. For example, Section 404 could have mandated that 75 percent of all wetlands would have to be saved. The law could have stipulated what types of wetlands were most important. Community Right-To-Know could have mandated that toxic chemical emissions be reduced by 50 percent over a five-year period. One advantage to these types of laws is that they are less dependent on active participation in the implementation phase. We can reach ecological outcomes without active community participation. A second advantage is that they would encourage a more open national discussion about what type of society citizens want to live in. Part of the problem with procedural legal systems is that they never really allow for a national discussion about important issues. Procedural laws mandate an open process. Over time, the process becomes the end goal. Rather than being proud of living responsibly, we become proud of having an open system.

The problem with outcome-based law, however, is that it depends upon a national debate during the drafting phase. Thus, we return to many of the original problems about information flows, ability to participate, and what views would be deemed legitimate. Outcome-based law is going to have a procedural component because the laws have to be drafted through some type of political process. But if grassroots groups are unable to access this political process (Grieder 1992), then the outcomes mandated by the laws are likely to reflect the interests of producers and developers.

Regardless of the option chosen, we find ourselves confronted with two basic dilemmas. First, either type of law is going to have a procedural component, and that procedural component has to be judged by how well it encourages participation. At both a theoretical and practical level, we have to get beyond the outcome/process dichotomy because it is a false one. Every procedural law has a clear outcome and must be publicly measured by how well it meets that outcome. Likewise, every outcome law is going to emerge from within a process and, thus, it is imperative that the process be open. Second, neither type of law is going to work unless we deal with issues of social stratification. We especially need to be cognizant of power inequalities. In most cases, legal processes do not nurture the context for a fair fight. Instead, grassroots groups are forced to fight on a grossly unequal playing field. We have a system based on competition taking place within an increasingly stratified society. My point is simply that law can only work when processes work. Neither processes of decision making nor processes of drafting outcomes can work fairly in highly stratified situations.

We probably lack the political will to redistribute power in any significant way. This shifts the burden more directly on state actors. The state has to be proactive in creating environments where disempowered groups can realistically battle more powerful social entities. The state has to come to see itself as in the business of helping communities mobilize against producers and developers, not because they wish to prejudice the outcome but, rather, because they believe in the process and want to create the conditions for process to work. The state has to play the role of advocate. In this role, the state has to ensure information flows, provide assistance to disempowered groups, and ensure that opinions are deemed legitimate. Under these conditions, it is possible that law can be crafted to support grassroots organizing.

ACKNOWLEDGMENTS

I would like to thank the following people for their help in preparing this paper: Allan Schnaiberg, Joan Mandle, Ken Gould, David Pellow, Wendy Espeland, Charles Ragin, Bob Nelson, Tamara Kay, Anne Weinberg, and Karen Haskin.

REFERENCES

Anderson, C. 1990. *Pragmatic Liberalism*. Chicago: University of Chicago Press.
Brown, P., and E. Mikkelsen. 1990. *No Safe Place*. Berkeley: University of California Press.
Cable, S., and C. Cable. 1995. *Environmental Problems, Grassroots Solutions: The Politics of Grassroots Environmental Conflict*. New York: St. Martin's Press.
Cable, S., and M. Benson. 1993. "Acting Locally." *Social Problems* 40: 464-477.
Cylke, K. 1993. *The Environment*. New York: Harper Collins.
Dunlap. R. 1992. "Trends in Public Opinion." In *American Environmentalism*, edited by R. Dunlap and A. Mertig. New York: Taylor & Francis.
Dunlap, R., and A. Mertig. 1992. "The Evolution of the U.S. Environmental Movement from 1970-1990: An Overview." Pp. 1-10 in *American Environmentalism*, edited by R. Dunlap and A. Mertig. New York: Taylor & Francis.
Edelstein, M. 1988. *Contaminated Communities*. Boulder, CO: Westview.
Freudenberg, N. 1984. *Not In Our Backyards!* New York: Monthly Review Press.
Freudenberg, N., and C. Steinsapir. 1992. "Not In Our Backyards: The Grassroots Environmental Movement." Pp. 235- 247 in *American Environmentalism*, edited by R. Dunlap and A. Mertig. New York: Taylor & Francis.

Friedman, L. 1975. *The Legal System: A Social Science Perspective.* New York: Russell Sage.

Geertz, C. 1983. *Local Knowledge.* New York: Basic Books.

Gould, K. 1991. "The Sweet Smell of Money: Economic Dependency and Local Environmental Political Mobilization." *Society and Natural Resources* 4: 133-150.

Gould, K. 1993. "Pollution and Perception: Social Visibility and Local Environmental Political Mobilization." *Qualitative Sociology* 16: 123-145.

Gould, K., A. Schnaiberg, and A. Weinberg. 1996. *Local Environmental Struggles: Citizen Activism in a Treadmill of Production.* New York: Cambridge University Press.

Greider, W. 1992. *Who Will Tell the People?* New York: Simon & Schuster.

Hadden, S. 1989. *A Citizen's Right To Know: Risk Communication and Public Policy.* Boulder, CO: Westview.

Harper, C. 1996. *Environment and Society.* Upper Saddle River, NJ: Prentice Hall.

Hawkins, K. 1984. *Environmental and Enforcement: Regulation and the Social Definition of Pollution.* Oxford: Oxford University Press.

Kennedy, D. 1976. "Form and Substance in Private Law Adjudication." *Harvard Law Review* 89: 1685-1778.

Kousis, M. 1991. "Development, Environment, and Mobilization: A Micro Level Analysis." *The Greek Review of Social Research* 80: 96-109.

Kousis, M. 1993. "Collective Resistance and Sustainable Development in Rural Greece: The Case of Geothermal Energy on the Island of Milos." *Sociologia Ruralis* 33: 132-146.

Landy, M., M. Roberts, and S. Thomas. 1990. *The Environmental Protection Agency: Asking the Wrong Question.* New York: Oxford University Press.

McAdams, D. 1982. *Political Process and the Development of Black Insurgency, 1930-1970.* Chicago: University of Chicago Press.

Mol, A. 1995. *The Refinement of Production.* Netherlands: International Books.

O'Connor, J. 1973. *The Fiscal Crisis of the State.* New York: St. Martin's Press.

O'Connor, J. 1988. "Capitalism, Nature, Socialism: A Theoretical Introduction." *Capitalism, Nature, Socialism* 1: 11-38.

Pellow, D. 1994. "Environmental Justice and Popular Epidemiology." Paper presented at the Annual Meetings of the American Sociological Association, Los Angeles, August 6-9.

Piven, F., and R. Cloward. 1979. *Poor People's Movements.* New York: Vintage Books.

Pring, G., and P. Canan. 1995. *SLAPPs: Getting Sued for Speaking Out.* Philadelphia, PA: Temple University Press.

Sale, K. 1993. *The Green Revolution: The American Environmental Movement 1962-1992.* New York: Hill and Wang.

Schnaiberg, A. 1980. *The Environment.* New York: Oxford University Press.

––––––. 1986. "Future Trajectories of Resource Distributional Conflicts." Pp. 435-444 in *Distributional Conflicts in Environmental-Resource Policy,* edited by A. Schnaiberg, N. Watts, and K. Zimmermann. New York: St. Martin's.

––––––. 1994. "The Political Economy of Environmental Problems and Policies: Consciousness, Conflict, and Control Capacity." Pp. 23-64 in *Advances in Human Ecology,* Vol 3, edited by L. Freese. Greenwich, CT: JAI Press.

Schnaiberg, A., and K. Gould. 1994. *Environment and Society.* New York: St. Martin's Press.

Shabecoff, P. 1993. *A Fierce Green Fire.* New York: Hill and Wang.

Simmons, J., and N. Stark. 1993. "Backyard Protest: Emergence, Expansion and Persistence of a Local Hazardous Waste Controversy." *The Policy Studies Journal* 21: 470-491.

Szasz, A. 1994. *Ecopopulism*. Minneapolis: University of Minnesota Press.

Williams, P. 1990. *The Alchemy of Race and Rights*. Cambridge, MA: Harvard University Press.

Yeager, P. 1991. *The Limits of Law*. New York: Cambridge University Press.

Advances in Human Ecology

J

A

I

Edited by **Lee Freese,** *Department of Sociology, Washington State University*

This series publishes theoretical, empirical, and review papers on scientific human ecology. Human ecology is interpreted to include structural and functional changes in human social organization and sociocultural systems as these changes may be affected by, interdependent with, or identical to changes in ecosystemic, evolutionary, or ethological processes, factors, or mechanisms. Three degrees of scope are included in this interpretation: (1) the adaptation of sociocultural forces to bioecological forces; (2) the interactions, or two-way adaptations, between sociocultural and bioecological forces; (3) the integration, or unified interactions, of sociocultural with bioecological forces.

P

R

The goal of the series is to promote the growth of human ecology as an interdisciplinary problem solving paradigm. Contributions are solicited without regard for particular theoretical, methodological, or disciplinary orthodoxies, and may range across ecological anthropology, ecological economics, ecological demography, ecological geography, biopolitics, and other relevant fields of specialization.

Volume 5, 1996, 303 pp. $78.50
ISBN 0-7623-0029-9

E

S

S

Also Available:
Volumes 1-4 (1992-1995) $73.25 each